Praise for *Bringing Nature Home*

"Reading this book will give you a new appreciation of the natural world—and how much wild creatures need gardens that mimic the disappearing wild."
—*The Minneapolis Star Tribune*

"An informative and engaging account of the ecological interactions between plants and wildlife, this fascinating handbook explains why exotic plants can hinder and confuse native creatures, from birds and bees to larger fauna."
—*Seattle Post-Intelligencer*

"A fascinating study of the trees, shrubs, and vines that feed the insects, birds, and other animals in the suburban garden."
—*The New York Times*

"Provides the rationale behind the use of native plants, a concept that has rapidly been gaining momentum. . . . The text makes a case for native plants and animals in a compelling and complete fashion."
—*The Washington Post*

"Tallamy explains eloquently how native plant species depend on native wildlife."
—*San Luis Obispo Tribune*

"[*Bringing Nature Home*] will persuade all of us to take a look at what is in our own yards with an eye to how we, too, can make a difference. It has already changed me."
—*Traverse City Record-Eagle*

"[*Bringing Nature Home*] delivers an important message for all gardeners: Choosing native plants fortifies birds and other wildlife and protects them from extinction."
—*WildBird* magazine

"A compelling argument for the use of native plants in gardens and landscapes."
—*Landscape Architecture*

"An essential guide for anyone interested in increasing biodiversity in the garden."
—*American Gardener*

Bringing Nature Home

HOW YOU CAN SUSTAIN WILDLIFE WITH NATIVE PLANTS

UPDATED AND EXPANDED

DOUGLAS W. TALLAMY

Foreword by Rick Darke

TIMBER PRESS
Portland, Oregon

Frontispiece: A mockingbird surveys his territory, making sure no other birds are eating the insects it contains.

First edition published in 2007.
Updated and expanded paperback edition published in 2009 by Timber Press.
Workman Publishing
Hachette Book Group, Inc.
1290 Avenue of the Americas
New York, New York 10104
timberpress.com

Eighteenth printing 2025
Printed in China on responsibly sourced paper

ISBN 978-0-88192-992-8

A catalog record for this book is available from the Library of Congress.

CONTENTS

FOREWORD

Once in a long while a book appears that fundamentally changes the way we think about our gardens and their role in the larger landscape. Provocative and powerfully persuasive, Doug Tallamy's *Bringing Nature Home* accomplishes this with grace and humor, blending solid ecological science and human social science to outline a modern recipe for inclusive habitat. Tallamy recognizes the changing dynamics of our world and suggests how individual gardeners, collectively, can protect and conserve the local biological diversity that is truly vital and irreplaceable.

Bringing Nature Home is a book many of us have been waiting for. So much more than a push for native plants, it articulates the broad interdependency of living relationships and literally redefines gardens as the new Nature. Tallamy tackles the potentially grim subjects of habitat destruction and resultant species loss and turns them into inspiring stories full of hope and opportunity. Drawing on a lifetime of experience as naturalist and observer, scientist and gardener Doug Tallamy writes with real familiarity about the sensory richness, vibrancy, and sustenance inherent in landscapes that are truly full of life.

Rich in concept and detail, this book asks and answers essential questions for modern gardeners inclined to good stewardship. How can we adjust our planting palette to be both beautiful and environmentally useful? How much more does a local oak species contribute to habitat richness than an out-of-ecological-context exotic tree? What do violets and fritillary butterflies, or pawpaws and zebra swallowtails have in common? Where might tomorrow's species come from? Spending some time with *Bringing Nature Home* and its wealth of revelatory moments is certain to enrich your understanding of how connected and contributing good gardens can be.

—Rick Darke

PREFACE

Occasionally we encounter a concept so obvious and intuitive that we have never thought to articulate it, so close to our noses that we could not see it, so entangled with our everyday experiences that we did not recognize it. In this book, I address several such concepts. Primarily, the wild creatures we enjoy and would like to have in our lives will not be here in the future if we take away their food and the places they live. I examine how we threaten their survival by trading our wild lands for uncontrolled expansion. And I emphasize the obvious consequence of that trade: in too many areas of our country there is no place left for wildlife but in the landscapes and gardens we ourselves create.

I also introduce ideas that are perhaps not so obvious. All plants are not created equal, particularly in their ability to support wildlife. Most of our native plant-eaters are not able to eat alien plants, and we are replacing native plants with alien species at an alarming rate, especially in the suburban gardens on which our wildlife increasingly depends. My central message is that unless we restore native plants to our suburban ecosystems, the future of biodiversity in the United States is dim.

Fortunately, two points of optimism temper this gloomy prediction. First and foremost, it is not yet too late to save most of the plants and animals that sustain the ecosystems on which we ourselves depend. Second, restoring native plants to most human-dominated landscapes is relatively easy to do.

Although I do suggest approaches here and there, this is not a how-to book; there are many other fine references on how to select and grow natives in different parts of the country. Nor is this a book about landscaping per se. I am not posing as a landscape architect, and I am not skilled in landscape design. I am simply proposing a justification for the liberal use of native plants in the landscape that has not yet been clearly articulated. I hope the reasoning presented in this book is logical and convincing, and maybe even entertaining.

I would like to thank my wife, Cindy, for keeping me on course in this and all other endeavors, as well as for her exceptional editorial skills and willingness to turn our property into a research station. For ideas, factual accuracy, advice, encouragement, and technical help, I also thank Tina Alban, Mary Ann Brown, Ed Bruno, Rick Darke, Vince D'Amico, Dale Hendricks, Bethany Plyler, Dot Plyler, and Jim Plyler, Kimberley Shropshire, Melinda Zoehrer, and all those I've met at conferences whose interest in these ideas has been my constant motivation.

CHAPTER ONE

Restoring Natives to Suburbia: A Call to Action

Gardeners enjoy their hobby for many reasons: a love of plants and nature, the satisfaction that comes from beautifying home and community, the pleasures of creative effort, the desire to collect rare or unusual species, and the healthful benefits of exercise and outdoor air. For some people, like my wife and me, there is pleasure in just watching plants grow.

But now, for the first time in its history, gardening has taken on a role that transcends the needs of the gardener. Like it or not, gardeners have become important players in the management of our nation's wildlife. It is now within the power of individual gardeners to do something that we all dream of doing: to make a difference. In this case, the "difference" will be to the future of biodiversity, to the native plants and animals of North America and the ecosystems that sustain them.

For decades, many horticulture writers have been pleading for a fresh appreciation of our American flora, and for almost as long they have been largely (or entirely) ignored. For several reasons, however, the day of the native ornamental is drawing near; the message is finally beginning to be heard. If I were to ask a random group of gardeners to comment on the importance of native plants in their gardens, they would probably recount several arguments that have been made in recent years in favor of natives over alien ornamentals. They might describe the "sense of place" that is created by using plants that "belong" or the dangers of releasing yet another species of invasive alien to outcompete and smother native vegetation. They might recognize the costly wastefulness of lawns populated with alien grasses that demand high-nitrogen fertilizers, broad-leaf herbicides, and pollution-belching mowers. Or they might mention the imperative of rescuing endangered native plants from extinction. These are all well-documented reasons for the increasing popularity of growing native plants.

Owners of native nurseries are also finding it easier and easier to enumerate the benefits of their offerings. Native plants are well adapted to their particular ecological niche and so are often far less difficult to grow than species from other altitudes, latitudes, and habitats. After all, these plants evolved here and were growing just fine long before we laid our heavy hands on the landscape.

Most compelling to me, however, is the use of native species to create simplified vestiges of the ecosystems that once made this land such a rich source of life for its indigenous peoples and, later, for European colonists and their descendants. That most of our ecosystems are no longer rich is beyond debate, and today, most of the surviving remnants of the native flora that formed them have been finished off by development or invaded by alien plant species. Too many Oak Parks, Hickory Hills, and Fox Hollows—developments named, as the environmentalist Bill McKibben has noted, for the bit of nature they have just extirpated—have been built across the country. Although relatively small, strategically placed and connected patches of completely restored habitats might foster the survival of some of our wildlife, I will describe later why such habitat islands can only protect a tiny fraction of the species that once thrived in North America. With 300 million human souls already present in the United States and no national recognition of the limits of our land's ability to support additional millions, we simply have not left enough intact habitat for most of our species to avoid extinction. All species need space in order to dodge the extinction bullet. So far we have not shared space very well with our fellow earthlings. In the following pages, I hope to convince you that, for our own good and certainly for the good of other species, we must do better. Native plants will play a disproportionately large role in our success.

The transition from alien ornamentals to native species will require a profound change in our perception of the landscaping value of native ornamentals. Europeans first fell in love with the exotic beauty of plants that evolved on other continents when the great explorers returned home with beautiful species no one had ever seen before. It quickly became fashionable and a signal of wealth and high status to landscape with alien ornamentals that no one else had access to. As the first foreign ornamentals became more common in the landscape, the motivation to seek new alien species increased. Even today, the drive to obtain unique species or cultivars is a primary factor governing how we select plants for our landscapes.

My epiphany

Although I chose entomology as a profession, I understand the thrill of growing an exotic plant for the first time. When I was in graduate school at the University of Maryland, I took a course in woody landscape plants from the noted horticulturist Robert Baker. He introduced me to the world of ornamental horticulture and the many alien species with landscape value. I left that course with an intense desire to plant as many of the species I had just learned about as possible. The only thing that slowed me down a bit was that I had no place to plant them. Still, I gathered seeds from many of the ornamentals on the University of Maryland campus, germinated them in the greenhouse, and planted the seedlings all over the yards of my parents and relatives. Among other things, my parents got a Japanese hardy orange, and I bestowed the gift of *Paulownia* trees, of all things, on unsuspecting Uncle George. I now find it ironic that, at the same time Robert Baker was turning me on to alien ornamentals, I was taking courses about plant-insect interactions. These were the courses that explained why most insect herbivores can only eat plants with which they share an evolutionary history. All of the information I needed to realize that covering the land with alien plant species might not be such a good idea had been neatly and simultaneously placed in my lap during those months in graduate school, but it was 20 years before I made the connection: our native insects will not be able to survive on alien plant species.

In 2000, my wife and I moved to 10 acres in southeastern Pennsylvania. The area had been farmed for centuries before being subdivided and sold

A view of our backyard shortly after we moved to southeastern Pennsylvania. The tangle of oriental bittersweet, multiflora rose, Japanese honeysuckle, and autumn olive—all alien species—is typical of what grows in so-called natural areas in the eastern United States.

to people like us who wanted a quiet rural setting close to work. We got our rural setting—sort of—but it was anything but the slice of nature we were seeking. Like many "open spaces" in this country, at least 35 percent of the vegetation on our property (yes, I measured it) consisted of aggressive plant species from other continents that were rapidly replacing what native plants we did have. We quickly agreed to make it a family goal to rid the property of alien plants and to replace them with the forest species that had evolved within the eastern deciduous biome over many millions of years. This rather optimistic and, I admit, peculiar use of our spare time has put us in intimate contact with the plants on our property, both alien and native, and with the wildlife that depends upon those plants.

Early on in my assault on the aliens in our yard, I noticed a rather striking pattern. The alien plants that were taking over the land—the multiflora roses, the autumn olives, the oriental bittersweets, the Japanese honeysuckles, the Bradford pears, the Norway maples, and the mile-a-minute weeds—all had very little or no leaf damage from insects, while the red maples, black and pin oaks, black cherries, black gums, black walnuts, and black willows

Alien plants like Bradford pear (A) and autumn olive (B) are avoided by native insects, while native plant species like black cherry (C) and red maple (D) are good food sources for native insect species.

had obviously supplied many insects with food. This was alarming because it suggested a consequence of the alien invasion occurring all over North America that neither I—nor anyone else, I discovered, after checking the scientific literature—had considered. If our native insect fauna cannot, or will not, use alien plants for food, then insect populations in areas with many alien plants will be smaller than insect populations in areas with all natives. This may sound like a gardener's dream: a land without insects! But because so many animals depend partially or entirely on insect protein for food, a land without insects is a land without most forms of higher life (Wilson 1987). Even the most incorrigible antienvironmentalist would be hard pressed to make an attractive case for such sterility. Pure anthropocentrists should be alarmed as well, since the terrestrial ecosystems on which we humans all depend for our own continued existence would cease to function without our six-legged friends.

But does the pattern of leaf damage I noticed in my backyard hold true? Does it occur elsewhere? If alien plants do reduce insect populations, by how much do they do so? Do aliens exclude all insect herbivores or just some?

The larva of *Actias luna*, the luna moth, is a beautiful member of the family Saturniidae that serves as an important source of food for birds, bats, and other creatures.

Are all alien plant species equally harmful to insects? And is the predicted effect on higher levels of the food web as serious as I've suggested? My colleagues and I have started the large, controlled research projects needed to address these important questions, and the data are starting to accumulate. So far, the results provide exciting support for gardeners who have already switched to natives or who are enthusiastic about doing so. If my concern for the fate of our insect herbivores turns out to be justified, these gardeners can and will "change the world" by changing what food is available for their local wildlife.

My argument for using native plant species moves beyond debatable values and ethics into the world of scientific fact. We can no longer hope to coexist with other animals if we continue to wage war on their homes and food supplies. This simple tenet provides an imperative, particularly for the bird and butterfly lovers among us, to fight invasive aliens as if it really matters and to reevaluate our centuries-old love affair with alien ornamentals. Beyond providing a challenge to ecologically minded gardeners, I will also explain how gardening with natives can create plantings that will stay beautiful and in balance without the use of pesticides. Gardening with natives is no longer just a peripheral option favored by vegetarians and erstwhile hippies. It is an important part of a paradigm shift in our shaky relationship with the planet that sustains us—one that mainstream gardeners can no longer afford to ignore.

CHAPTER TWO

The Vital New Role of the Suburban Garden

I needn't elaborate on the many things our gardens do for us. Properly designed, gardens tie our homes to the surrounding landscape as well as provide an outlet for artistic expression and a source of natural beauty that can be enjoyed year round. Our gardens can also offer refuge from an increasingly hectic and unpleasant world. But because gardens are, in essence, groups of plants, they also have the potential to perform the same essential biological roles fulfilled by healthy plant communities everywhere.

Plants are Earth's lifeblood

Plants are not optional on this planet. With few exceptions, neither we, nor anything else, can live without them. We invariably take plants and the bene-

Plants are the fundamental source of energy for all terrestrial creatures.

Animal diversity is high in the tropics because plant diversity is high.

fits they provide for granted. Who takes time to think that the oxygen in each breath we take has been produced exclusively by plants? Who is grateful for the forests when we are blessed with the rains that provide the fresh water we all require, water that is filtered clean by the tangled mass of roots it flows through en route to the nearest stream? Even farther from our consciousness is the primary role of plants in the food chain (more accurately, a wonderfully intricate food web). Nearly every creature on this planet owes its existence to plants, the only organisms capable of capturing the sun's energy and, through photosynthesis, turning that energy into food for the rest of us. Only in the deepest reaches of the ocean do life forms survive that don't require this food, deriving their energy through chemosynthesis of sulfur from deep-sea vents (Ruby, Wirsen & Jannasch 1981). Plants, therefore, form the first trophic level: the energy that sustains all life.

Because animals directly or indirectly depend on plants for their food, the diversity of animals in a particular habitat is very closely linked to the diversity of the plants in that habitat (Rosenzweig 1995). When there are many species of plants, there are many species of animals. Because plants are so different from one another in their size, shape, habit, their soil, water, and nutrient requirements, and their leaf chemistry (the most important factor determining taste), greater numbers of plant species mean more opportunities for animals to obtain their energy without interfering with one another. That is, plant diversity creates niches to which animals adapt over evolutionary time. This is why we hear so much about the incredible animal diversity of the tropics. There are so many different types of animals in tropical ecosystems because plant diversity is so high there. For example, a single hectare (2.47 acres) of Amazonian rainforest in Ecuador can support as many as 473 species of trees (Valencia, Balslev & Paz y Mino 1994), whereas there are only 134 species of trees in all of Pennsylvania (Rhoads & Block 2005). So if we want to create ecosystems with a diversity of animal species, we first have to encourage a healthy diversity of plants.

Why insects are essential

The second trophic level comprises all the animals that eat plants: the herbivores, or phytophages. In our neck of the woods, the most familiar and apparent herbivores are white-tailed deer, rabbits, and groundhogs. My wife and I were reminded of the strict herbivory of beavers when one showed up in our neighbors' pond and made meals of their birch and willow trees. Other common vertebrates, such as chipmunks, squirrels, mice, raccoons, box turtles, and of course humans, include plants in their diets but are not restricted to them. Many of these omnivorous creatures are relatively large, and most fall into the category of what have been termed "charismatic megafauna." It may be a surprise that when it comes to transferring energy from the first trophic level (plants) to the predators, parasites, and omnivores in other trophic levels, these charismatic vertebrates are relatively unimportant. What, then, do most animals in higher trophic levels rely on to pass on the energy held within the plant? Insects!

I cannot overemphasize how important insect herbivores are to the health of all terrestrial ecosystems. Worldwide, 37 percent of animal species are herbivorous insects (Wiess & Berenbaum 1988). These species are collectively very good at converting plant tissue of all types to insect tissue, and as a consequence they also excel at providing food—in the form of themselves—for other species. In fact, a large percentage of the world's fauna depends entirely on insects to access the energy stored in plants (Wilson 1987). Birds are a particularly good example of such organisms. If you count all of the terrestrial bird species in North America that rely on insects and other arthropods (typically, the spiders that eat insects) to feed their young,

Rabbits are a good example of a mammalian member of the second trophic level, an animal that eats only plants.

Omnivores like the box turtle eat both plants and other animals.

Nearly all terrestrial birds rear their young on insects, not seeds or berries. The bluebird nourishes its young with herbivorous insects that have captured the energy stored by plants.

Insects transfer the most energy from plants to animals in higher trophic levels. Juicy caterpillars like the spiny oak slug (*Euclea delphinii*) provide valuable nutrients and energy for birds and other animals.

The typical suburban landscape is a highly simplified community consisting of a few species of alien ornamental plants that provide neither food nor shelter for animals. Our challenge is to redesign our living spaces in ways that provide both.

you would find that figure to be about 96 percent (Dickinson 1999)—in other words, nearly all of them.

And no wonder! Insects are unusually nutritious. Pound for pound, most insect species contain more protein than beef, and their bodies are extremely high in valuable energy (Lyon 1996). The Pulitzer prize–winning author and renowned ecologist E. O. Wilson (1987) has called insects "the little things that run the world," in part because of their role in transferring energy from plants to other animals that cannot eat plants directly. In sum, if we want to have members of higher trophic levels in our managed ecosystems, we must also have their primary food source: insects.

* * *

It is increasingly clear, as we shall see, that much of our wildlife will not be able to survive unless food, shelter, and nest sites can be found in suburban habitats. Let's focus on the first of these essential resources: food. Because food for all animals starts with the energy harnessed by plants, the plants we grow in our gardens have the critical role of sustaining, directly or indirectly, all of the animals with which we share our living spaces. The degree to which the plants in our gardens succeed in this regard will determine the diversity and numbers of wildlife that can survive in managed landscapes. And because it is we who decide what plants will grow in our gardens, the responsibility for our nation's biodiversity lies largely with us. Which animals will make it and which will not? We help make this decision every time we plant or remove something from our yards.

Unfortunately, because we have been so slow to recognize the unprecedented importance of suburban gardens for the preservation of wildlife, gardeners across the nation have been caught off guard. We have proceeded with garden design as we always have, with no knowledge of the new role our gardens play—and, alas, it shows. All too often the first step in the suburbanization of an area is to bulldoze the plant assemblages native to our neighborhoods and then to replace them with large manicured lawns bordered by a relatively few species of popular ornamentals from other continents. Throughout suburbia, we have decimated the native plant diversity that historically supported our favorite birds and mammals.

CHAPTER THREE

No Place to Hide

When I was nine years old, I learned firsthand about the finality of suburban development as practiced today. Having recently moved with my parents and siblings to Berkeley Heights, New Jersey, I spent my summer days exploring the "wild" places that surrounded me. One of my first discoveries was a small pond crowded with life right next door to my house. Thousands of pollywogs wiggled near its shoreline, and I took great delight each day in watching them grow their little legs—first the hind legs, then the forelegs, millimeter by millimeter. Eventually, the first minute toads hopped out of the water and into the weeds, tiny replicas of their parents, whose mating song I had learned to imitate the previous spring. As I watched the little guys jump about, a bulldozer crested nearby piles of dirt, and in an act that has been replicated around the nation millions of times since, proceeded to bury the young toads and all of the other living treasures within the pond. I might have been buried too, if I hadn't given up trying to rescue the toads. I saved about 10 that day, but for nothing: the pond was gone, leaving nowhere for the toads to breed. Within two years, a toad was a rare sight near my house; soon they were completely gone, along with the garter snakes, whose main prey they had been, and other members of the food web supported by the life in that pond. I had witnessed the local extinction of a thriving community of animals, sacrificed so that my neighbors-to-be could have an expansive

To build our sprawling suburbs, we have indiscriminately destroyed the breeding habitats of countless animals like the American toad.

lawn. (Those neighbors would eventually pay me two dollars a week to mow that new lawn.)

Humans versus nature

It is easy to understand the conflict that arises between humans and nature as human populations grow. We bring to every encounter with nature an ancient struggle for our own survival. In the old days, all too often it was nature—her predators, winters, floods, and droughts—that did us in. Those of our ancestors who were particularly good at conquering nature were the ones who survived and reproduced, so we all share genes that encourage us to beat back nature at every turn. We have become accustomed to meeting our needs without compromise. If we need space to live, we take it—all of it—and if that means filling in a pollywog pond or cutting down a woodlot, then so be it. We feel completely justified in sending the plants and animals that depend on those habitats off to make do someplace else. This is partly because no one is going to choose a pollywog over a human if presented with such a choice, and partly because, until recently, there always has been someplace else for nature to thrive.

But no longer. We can no longer safely relegate nature to our parks and preserves, assured that it will be there for us when we need it. We can no longer replace the native vegetation in our neighborhoods with foreign plants and remain confident that our native species will survive somewhere else. We can no longer rely on local natural areas to supply food and shelter to the birds, mammals, reptiles, and amphibians of North America.

Why we can no longer take nature's persistence for granted becomes evident when we consider how much of the landscape we have taken for our own use. We humans have co-opted such a large percentage of natural areas that, in far too much of the country, there are no undisturbed habitats left (Rosenzweig 2003). What's more, plants from other continents have now invaded the tiny remnants of the great ecosystems that once sustained our biodiversity (Mooney & Hobbs 2000). As we take more and more space for our own use, we are consigning the animals that used to live there to "natural" areas in which many of the plants that constitute the vital first trophic level are no longer "natural": they are alien species that have either been planted deliberately or have escaped from our gardens, ports, or rail yards. There simply are not enough native plants left in the "wild"—that is, not enough undisturbed habitat remaining in the United States—to support the diversity of wildlife most of us would like to see survive into the distant future.

The drivers of diversity

To understand why we need to restore the ecological integrity of suburbia in order to prevent the extinction of most of our plants and animals, we must first understand what creates and maintains diversity. If we look at species diversity around the world, the first thing we notice is that the number of species in a given area depends on the size of the area (Rosenzweig 2003). Large continents have more species than small continents, and continents have many more species than islands. There are two reasons for this relationship between area and species number. First, the rate at which new species are created is higher in large areas than in small areas. This is because the primary cause of speciation, the geographic isolation of some individuals from other individuals of the same species, is far more likely to occur in large places than in small places (Mayr 1942). Second, the rate at which species go extinct is slower on large land masses than on small ones (Dobson 1996). The huge size of continents, for example, allows species to occupy large ranges, reducing the likelihood that a hurricane, cold snap, drought, or volcano will kill all members of a species. We learned long ago from Robert MacArthur and E. O. Wilson (1967) that the number of species existing in any one place represents a balance—an equilibrium, as they put it—that is reached over time between the rate at which new species arrive through speciation and immigration and the rate at which existing species disappear through extinction and emigration. And so, because species form faster and disappear slower on continents, we find the greatest number of species on these huge land masses.

Habitat fragmentation equals extinction

Problems for biodiversity in North America started when humans began destroying the diverse forests and grasslands that once covered the continent in order to plant crops and create living spaces that resemble the savannah parkland in which our species feels most comfortable. We did not systematically start at one end of the continent and wipe out everything as we proceeded. Instead, we left islands of suitable habitat in which most of the plants and animals that survive today found refuge. At first, these habitat patches were relatively large, but today they are miniscule, far too small to sustain populations of most living things for very long. For example, as of 2002, the once contiguous forest cover of Delaware had been reduced to 23 percent of its original size; 46 percent of the small woodlots that remain are less than 10 acres in size. It is curious that the news media have drawn our attention to the loss of tropical forests yet have been silent when it comes to how we have devastated our own forests here in the temperate zone. Only 15 percent of the Amazonian basin has been logged, whereas well over 70 percent of the forests along our eastern seaboard are gone (Brown 2006). We have reduced the enormous land mass that, over millions of years, created the rich biodiversity we can still see today in this country to tiny habitat islands. And therein lies the problem. Tiny habitat islands have high rates of species extinction and emigration and low rates of speciation and immigration.

Immigration? Just where are animals supposed to immigrate from? There are no large, pristine, and productive areas left to provide a pool of individuals for immigration to our habitat islands. We have only what ecologists call "sink" habitats: isolated locales that constantly lose individuals to death and emigration. As the "extinction debt" of the habitat islands we have created comes due, we will end up with far fewer plant and animal species. Fortunately, extinction takes a while, but because we have already fragmented the continent into habitat islands, we have set the clock ticking for our biodiversity and time is running out.

An excellent example of how habitat fragmentation leads to local extinction over time is currently being documented on Barro Colorado Island, an island in the middle of the Panama Canal (Robinson 1999). In 1914, this patch of tropical forest was the top of a small mountain rather than an island and was surrounded by hundreds of square miles of undisturbed habitat. When the canal was completed, the Chagres River was dammed, and the rising waters of what is now called Gatun Lake turned the mountaintop into an island isolated from the adjacent forest. Barro Colorado Island may not seem that isolated to us; the distance to the nearest shore in most places is less than a mile. But to most of its animals and plants, it may as well be in the middle of the Atlantic Ocean. From the start, its isolation from the surrounding forest began to take a toll on the number and kinds of animals that would survive on the island. Because most forest animals are reluctant to leave their habitat, the island's woodland creatures were cut off from any interactions with the mainland. They would not swim or fly to find food or mates or to disperse to new territories. Even though the island spreads over 3700 acres, a sizable chunk of real estate compared to most of the habitat islands we have created in North America, it is too small to sustain the populations of many of its original inhabitants. Sixty-five species of birds have disappeared from the island since 1914, and many others are on the brink of extinction. In all the years since Barro Colorado Island was cut off from the mainland, the extinction debt has only been partially paid. No one knows how long it will take before the island community reaches its final equilibrium number of species balanced between extinction and immigration. What is clear, however, is that the number of species that can be sustained on the island is a fraction of what existed on that same acreage at the beginning of the 20th century.

Area effects are no less devastating on habitat islands—forests surrounded by developments rather than by water—than they are on real islands. The creatures that depend on the resources found only in forests cannot make a go of it among manicured lawns crisscrossed by paved roads. Just as on islands surrounded by water, the loss of species from habitat islands takes time, but it inevitably happens. For example, Ashdown Forest in Sussex, England, the inspirational setting of A. A. Milne's *Winnie-the-Pooh*, has become entirely isolated by development. Since the 1920s, when Milne was writing his famous children's stories, the forest has lost 47 plant species (Marren 2001).

What do Barro Colorado Island and Ashdown Forest tell us about the number of species we can expect to survive habitat fragmentation in the United States? You might think this would be impossible to predict accurately, but studies by Michael Rosenzweig and his students at the University of Arizona (1995, 2003) have shown that such predictions are as simple as can be. As we have just seen, species diversity is a function of the area of suitable habitat that is available for plants and animals, as well as the time it takes to reach species equilibrium in that habitat. If you turn the clock forward to the point at which this equilibrium has been reached, you will find that the number of species that will survive human habitat destruction is a simple percentage of the amount of habitat we leave undisturbed, a 1:1 correspondence. For example, if we take 50 percent of the land in the United States for our own use, we will end up with 50 percent of the species that originally inhabited this country. If we usurp 80 percent of the land, we will lose 80 percent of the species. So the mystery is gone. We now know exactly how our actions are going to affect biodiversity if we continue on our present course.

The simple relationship between species survival and habitat area gives real urgency to the next question. How much land have we already taken? This is such a big country. Surely there is plenty of undisturbed habitat out there . . . somewhere. Our perception has always been that, no matter how many subdivisions we build, or how much land we put to the plow, or how many roads we construct, there will always be plenty of undisturbed space left. As long as we believe this, there is no reason to reevaluate our use of the land. This complacent view is not one held only by a few old-timers who remember the open, undisturbed spaces of their youth. I once asked one of my 20-year-old students how much land he thought humans had taken in the United States. Bless his optimistic heart: without hesitation he estimated that we have modified just 20 percent of the land for our own use. But we need not guess at this critical figure. Global Information Systems and satellite images now allow us to make very precise measurements of land use anywhere in the world.

We have taken it all

Let's start by looking at where we live. According to the USDA Economic Research Service, over 60 percent of the U.S. population lived in rural areas in 1900. This number had decreased to 36 percent by 1950 and sank to a mere 17.4 percent in 2000. Today, almost 83 percent of our 300 million people live either in cities or in sprawling suburbs (U.S. Census Bureau 2005). By 1986, we had converted over 69 million acres to managed urban and suburban landscapes (Grey & Deneke 1986). That is an area 53 times the size of Delaware. In Pennsylvania, my home state, less than 1 percent of the land can still be considered "wild" (Coleman 2003). According to the USDA's Natural Resource Conservation Service, 2 million acres—an area the size of Yellowstone National Park—were lost to development nationally each year between 1982 and 1997, and the rate of development has been accelerating (McKinney 2002). Incredibly, suburbia in some areas of the country has increased 5909 percent since 1960 (Hayden 2004).

Unfortunately, that's not the end of the grim news. We have paved at least 4 million linear miles of public roads in this country. (Hayden 2004). Add parking lots, driveways, and other paved surfaces to our streets and highways, and you have 43,480 square miles of blacktop smeared over the lower 48 states (Elvidge et al. 2004)—an area five and a half times the size of New Jersey. Regrettably, our rate of paving is increasing. Impervious surface increased more than 40 percent in the Chesapeake Bay watershed between 1990 and 2000 (They paved paradise, 2004).

Second only to paving in its impact on biodiversity is our love affair with sterile lawns. Christina Milesi has estimated that we have converted between 32 and 40 million acres—as much as 62,500 square miles—to suburban lawns in this country (Milesi et al. 2005). That is an area more than eight times the size of New Jersey dedicated to alien grasses! For further confirmation of our urbanization of North America, glance at the composite image derived from hundreds of photos taken by the U.S. Defense Meteorological Satellites Program (DMSP) in the mid-1990s. Each spot of light represents a town, city, or major metropolitan center, all of which are surrounded by miles and miles of suburbia. Huge areas east of the Mississippi have been completely "developed."

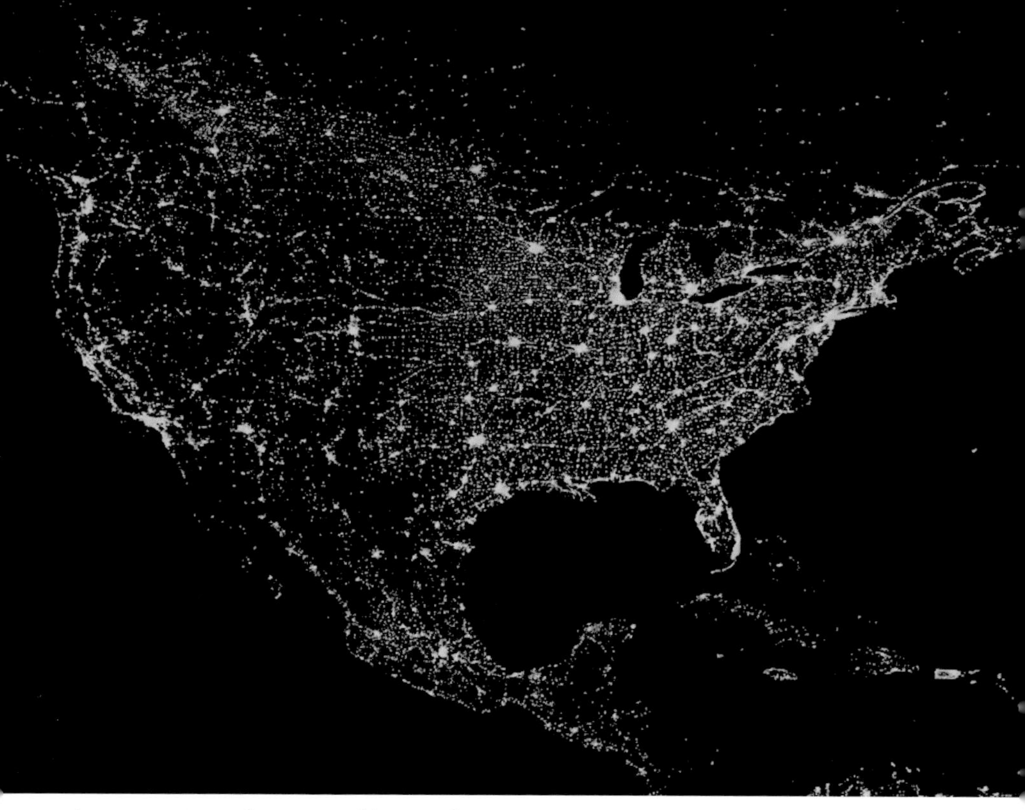

A composite image from space of the United States at night shows the extent to which we have converted natural areas to developed landscapes. Image from NASA/DMSP.

The landscape of the United States is crisscrossed with over 4 million miles of paved roadways.

Why local extinction is a local crisis

As you might have guessed, the amphibians in my pollywog pond are not the only animals whose populations have been decimated by rampant development. As of 2002, Delaware had lost 78 percent of its freshwater mussel species, 34 percent of its dragonflies, 20 percent of its fish species, and 31 percent of its reptiles and amphibians. Moreover, 40 percent of all native plant species in Delaware are threatened or already extirpated from the state, and 41 percent of Delaware's bird species that depend on forest cover are now rare or absent (Vargo & Gallagher 2002; Hess et al. 2000). In fact, we know most about the effects of habitat loss from studies of birds. For those who would like birds in their future, the statistics are truly frightening. Neotropical migrants, such as wood thrushes, warblers, catbirds, hawks, wrens, vireos, flycatchers, kingbirds, nightjars, swallows, tanagers, orioles—species that fly thousands of miles to Central or South America to spend the winter—have declined an average of 1 percent per year since 1966 (Sauer, Hines & Fallon 2005). Add up those percentages, and you're looking at nearly a 50 percent reduction in population sizes for many of our bird species within the space of 50 years.

Habitat loss is not only a problem in the United States; it has put birds under siege around the globe. The World Conservation Union has estimated that 12 percent of all bird species are threatened with extinction because of habitat loss and invasive species. Conservation ecologists such as Cagan Sekercioglu of Stanford University believe that one fourth of all bird species will be functionally extinct (that is, so rare that they no longer contribute to the function of ecosystems) within a century (Sekercioglu, Daily & Ehrlich 2004).

We are losing our birds because we have taken away their homes and their food and filled their world with dangerous obstacles that take a terrible toll. Remember those 4 million miles of U.S. roads? Vehicles using them flatten birds daily, resulting in 50 to 100 million deaths per year (U.S. Fish & Wildlife Service 2002). Then there are buildings along the roads—often tall and always window-filled buildings. Each year as many as 1 billion birds are killed when they fly into the windows of those buildings (Klem 1990). I

(right) Neotropical migrants like the wood thrush (top) have been hit particularly hard by habitat loss, but even the breeding habitat for common birds like the black-capped chickadee (below) is disappearing quickly. Wood thrush photo by William P. Brown.

won't go into the numbers of birds killed by cats that live in and around these buildings, or the number that fly into cell towers standing near the buildings. Naysayers like Dixy Lee Ray (Ray & Guzzo 1994) and Ronald Bailey (1993) have countered those who express alarm at such findings with their own statistics. The United States has more birds than ever before, they claim. What they don't mention is that the birds they are counting are European starlings and house sparrows, both invasive species from European cities. Scientific consensus is that our native birds are in deep trouble, and we are going to have to improve their habitats quickly if they are to survive at all.

An alarming prediction

I have buried you under a depressing pile of statistics on the amount of land humans have converted for their own use in this country, and I haven't even yet mentioned how much land we have left undisturbed. Ecologists have worked hard since the mid-1990s to come up with this figure. Their estimates depend on how strictly you define "undisturbed," but the consensus among landscape ecologists is that 3 to 5 percent of the land remains as undisturbed habitat for plants and animals (Rosenzweig 2003). In other words, we have taken and modified for our own use between 95 and 97 percent of all land in the lower 48 states. The 2002 USDA Census of Agriculture tells us that 41.4 percent of our land is in agriculture, which means that we have converted 53.6 to 55.6 percent of the land to cities and suburbia. As far as our wildlife is concerned, we have shrunk the continental United States to 1⁄20 its original size. And because our refuges and woodlots are not contiguous habitats, but survive as scattered islands from coast to coast, the effective size of undisturbed land in the United States is far smaller than those statistics indicate. When extinction adjusts the number of species to the land area that remains for the plants, mammals, reptiles, birds, and invertebrates of North America (something that will happen within most of our lifetimes), we will have lost 95 percent of the species that greeted the Pilgrims.

Unless we modify the places we live, work, and play to meet not only our own needs but the needs of other species as well, nearly all species of wildlife native to the United States will disappear forever. This is not speculation. It is a prediction backed by decades of research on species-area relationships by ecologists who know of what they speak. And the extinction of our plants and animals is not a scenario lost in the distant future. It is playing out across the country and the planet as I write. Our preserves and national parks are not adequate to prevent the predicted loss of species, and we have run out of the space required to make them big enough. For conservationists, and indeed for anyone who celebrates life on earth, this is perhaps the direst possible consequence of the human enterprise.

An easy solution

There is, however, a way out of this mess—*if* we act before the extinction debt we have created is realized. The predictions of mass extinction are based on the assumption that the vast majority of plants and animals cannot coexist with humans in the same place at the same time. Nonsense! Evidence suggests that the opposite is true: most species could live quite nicely with humans if their most basic ecological needs were met. Yes, some species such as the cougar, gray wolf, and ivory-billed woodpecker are just too reclusive to become our fellows. But countless others could live sustainably with us if we would just design our living spaces to accommodate them. In too many places, we have removed the food, shelter, and nesting sites needed by most species in our haste to make vast parking lots and shopping malls, lawns and soccer fields. In no place, however, does it have to be that way. We have excluded other species from our living spaces through thoughtlessness, not through need. It is not only possible but highly desirable from a human perspective to create living spaces that are themselves functioning, sustainable ecosystems with high species diversity.

Our impact on every square mile of the earth's biosphere, that thin zone on the planet's surface in which the conditions for life are ideal, has been so great that we must give up the old notion of preserving nature in its pristine form. That, however, does not mean that we must give up on nature altogether. Nature's living components—its 9 million or so species of plants and animals—are, for the most part, still with us, although most species are in a desperate struggle to adapt to the changes we impose on their environment daily. Michael Rosenzweig (2003) calls the redesign of human habitats for the accommodation of other species "reconciliation ecology." It's a mouthful, but it's the future; and we will play a central role in its success simply by reevaluating our use of native plants in the landscapes in which we live.

CHAPTER FOUR

Who Cares about Biodiversity?

My father once asked me, "What good is a house fly? What is its purpose?" His question reflected the Judeo-Christian belief that the earth and all of its creatures were created to serve humans in some way. Because I was a graduate student in entomology at the time, he thought I should have learned how house flies meet human needs, something that wasn't obvious to him. It wasn't obvious to me either, so I dodged his question by posing one of my own: "What good are humans, and what purpose do *they* serve?" It is still not clear to me how house flies, or polar bears, deep-sea tube worms, the polio virus, or the human bot fly, serve humans. Though fun (or scary) to think about, such questions quickly move beyond hypothesis testing—the backbone of science—to religion and philosophy and therefore are never resolved beyond personal convictions.

If we change the question just a little, however, science can weigh in on the subject with data from carefully designed, repeatable experiments. Instead of asking about the "purpose" of any given species, let's ask, "What role does a particular species play within its ecosystem?" or "What good is biodiversity?" or, even better, "How do diverse ecosystems differ from simplified ecosystems?" We can even yield to our anthropocentric tendencies and ask, "How does the quality of human life differ in simple and diverse ecosystems?" The problem with recasting the question in this way, of course, is that environmentalists are constantly challenged by nonenvironmentalists to justify preserving biodiversity. Are plants and animals other than those manipulated by agriculture really necessary? After all, millions of people live quite successfully in New York City, an environment almost devoid of non-human life forms.

Why we need other forms of life

I have heard four answers to queries about the need for biodiversity. The first is that we don't need it. In fact, many people would argue that efforts to preserve biodiversity have caused nothing but roadblocks and headaches for economic development. The second answer is that we do need biodiversity so that we can exploit it, and the example we hear most is that we might find new medicines or pesticides in plants and animals if we preserve rainforests. Third, it has been argued that biodiversity should be preserved for aesthetic, ethical, or moral reasons, that there is no justification for our wanton destruction of other species. Indeed some advocates, E. O. Wilson, for example, believe that our desire to save the natural world lies in our biophilia, our innate love of nature (Wilson 2002). Finally, an argument that is only now getting its share of ink is that we need biodiversity because it literally sustains us. The three arguments in favor of preserving biodiversity are not mutually exclusive; we can preserve the plants and animals that remain on this planet so that we can use them, because we like them, and because we absolutely need them for our own survival. Only the first argument—that we do not need biodiversity at all—is at odds with the other views. So which is it? Is biodiversity necessary or not? Are there measurable consequences, beyond a sense of loss, from reducing biodiversity, and, if so, what are they?

The rivet analogy

One of the earliest attempts to justify complex over simple ecosystems was a model advanced by Paul and Anne Ehrlich in 1981. The Ehrlichs likened an ecosystem to an airplane in flight, and the species within the ecosystem to the rivets holding the plane together. They pointed out that a plane is able to fly because its thousands of parts are held together by rivets in a way that allows them to function as a cohesive unit, resulting in a machine that can defy gravity. Each part, especially the essential ones like the wings and engines, is fastened to the plane's body with lots of rivets. If one or two rivets pop and are lost, there are enough remaining rivets to keep the part from falling off, and the plane will still fly. But if enough rivets are lost, a part vital to the plane's ability to stay in the air will stop working, and the plane will become crippled and crash.

The Ehrlichs' plane image was effective because we could visualize the consequence of lost rivets all too well. It also introduced the concept of redundancy—several rivets working together to perform the same task—to hold the wing to the body, for example. In a diverse ecosystem, many species perform similar tasks. Penstemon flowers, for instance, might be visited by three species of bumblebees, five species of moths, and one hummingbird species. If one or two of those pollinators disappear, the plants will still be pollinated and make viable seeds. The rodents that eat those seeds will still have food, as will the screech owls that eat the rodents. Redundancy in pollinators will save the day. But if the ecosystem is depauperate in pollinating species, the loss of one pollinator might quickly lead to the local extinction of the penstemon population. This would not only hurt the rodents and the screech owls, but it would also lead to the extinction of all insects that eat only penstemon leaves. These losses would then reduce the food available for all insectivores that included those insect specialists in their own diets. If this food reduction were large enough, insectivorous birds that bred in the area would not be able to feed their young, and they too would be lost from the ecosystem. These types of connections are not far-fetched; they are common in all ecosystems and are just as significant as the interconnected parts that keep planes in the air.

The role of keystone species

The Ehrlichs thus described ecosystems pretty well. Most ecosystems are highly complex and consist of many species that rely on the presence of many other species to make a living. Moreover, redundancy seems to be a common feature of ecosystems. But the airplane metaphor leaves one important question unanswered. How many species can be lost from an ecosystem before it crashes? That is, just how much diversity is necessary to maintain ecosystem functions? Simple manipulations performed by Robert T. Paine in 1969 further complicated this question. Paine studied species interactions in the rocky tidal pools of the Pacific Coast. He found that removing some spe-

Beavers have been described as keystone species because their removal from an ecosystem threatens the survival of other species within that system (Power et al. 1996).

cies from a pool had little effect on the populations of the remaining species. But if he removed *Pisaster ochraceus,* a predatory starfish, the community collapsed, with the loss of nearly half of the species from the pools. Without its primary predator, the mussel *Mytilus trossulus* became so numerous that it excluded other species from the tidal pools.

Paine coined the term "keystone species" to describe the essential role filled by *Pisaster ochraceus* in the intertidal community and in the maintenance of its diversity and species composition. The clear implication of Paine's work was that some species are more important to the sustainability of ecosystems than others—or, in the Ehrlichs' terms, some rivets in our hypothetical plane are more critical to the plane's function than others. I wouldn't mind much if a rivet fastening a chair to the floor popped, but if rivets holding an engine in place started popping, my blood pressure might pop as well.

Ever since Paine's seminal work was published, ecologists worldwide have been trying to identify keystone species. They have found many. Black-tailed prairie dogs, sea otters, African elephants, cottonwoods, pumas, beavers, and even the extinct dodo bird are just a few of the species that have (or had) keystone roles in their ecosystems. From the perspective of the conservationist, saving keystone species first became paramount because by doing so, many other species would be saved as well. But conservationists soon faced a problem. Too many species were turning out to have keystone roles in their ecosystems. In fact, so many species have been found to control the fate of other species that some scientists have recently begun to question the usefulness of the keystone concept altogether. Even more problematic was the realization that in many systems, a particular species seems to play a keystone role in maintaining ecosystem diversity and function only some of the time. At other times, removing that species has little or no measurable affect on the ecosystem. Could Paine have been wrong after all?

The Jenga hypothesis

In teasing out an answer, we must remember that the ecosystem Paine studied in 1969 comprised relatively few species, and growing evidence suggests that certain species are more likely to play keystone roles in simple ecosystems than in more diverse ones. It does not follow, however, that there are no central players in more complex ecosystems. Peter de Ruiter and his colleagues at Utrecht University think that the misconception starts when we view an ecosystem as a static entity with fixed food linkages among species (de Ruiter et al. 2005). They suggest instead that we should liken ecosystems to the game known as Jenga. In Jenga, blocks are used to build a free-standing tower. Once the tower has been constructed, the goal of the game is for players to remove blocks from the tower, one at a time, without causing the tower's collapse. In the Jenga metaphor, the role of any given species in maintaining the stability of its ecosystem is similar to the role individual blocks play in keeping the tower from tumbling down. Each block (except those in the topmost layer) supports the tower in some way. How vital that support is depends on which other blocks are present in the tower at any given time. Every time a block is removed, the relative importance of the remaining blocks changes. Thus, the role each block plays in the stability of the tower is relative and constantly changing. If ecosystems are Jenga towers, almost any species can play a keystone role under the appropriate circumstances.

I find the Jenga metaphor compelling for several reasons. First, it explains why we find so many species playing keystone roles, albeit at varying times. Ecosystems are not simple arches that depend upon the presence of a single keystone for their existence. Rather, they are complex buildings that stand successfully because of the support of many load-bearing girders. De Ruiter's Jenga metaphor improves on the Ehrlichs' airplane by enabling us to appreciate the potential importance of all species to the continued existence of an ecosystem. We have no trouble locating the critical rivets in an airplane because we know which ones support vital structures. But in a Jenga tower this becomes very challenging. We are never certain beforehand whether a particular block is holding up the tower or whether it can be removed without effect.

If research continues to support the notion that ecosystems are Jenga towers rather than airplanes, conservation efforts will actually become more straightforward than they were under keystone models. With the Jenga approach, we no longer have to worry about correctly identifying the keystone species in an ecosystem, and we no longer have to fret over the sacrifice of species that do not play an obvious keystone role—*all* species have the potential to sink or save the ecosystem, depending on the circumstances. Knowing that we must preserve ecosystems with as many of their interacting species as possible defines our challenge in no uncertain terms. It helps us to focus on the ecosystem as an integrated functioning unit, and it deemphasizes the conservation of single species. Surely this more comprehensive approach is the way to go.

Diversity and ecosystem health

All of this suggests that biodiversity is essential to the stability—indeed, the very existence—of most ecosystems. We remove species from our nation's ecosystems at the risk of their complete collapse. But biodiversity also plays an important role in the efficiency with which ecosystems function (Kinzig, Pacala & Tilman 2002). Ecosystem efficiency can be determined by measuring how long energy is retained in an ecosystem before being lost. Not all species use energy with equal efficiency. When a community has many different species, the chance that some of those species will be efficient energy-users increases. Energy flowing through ecosystems with many types of species is therefore used more efficiently, and with less loss to the surrounding environment, than energy entering simplified ecosystems. More energy in the system means that the system will be more productive (it will produce more plant and animal biomass, that is, weight) and, from a selfish human perspective, produce more ecosystem services for us (make more fish, more lumber, and more oxygen, filter more water, sequester more carbon dioxide, buffer larger weather systems, and so on).

Biodiversity also benefits ecosystems by making them less susceptible to alien invaders (Kennedy et al. 2002). An ecosystem that is highly diverse has more of its available niches filled by competing organisms and therefore is able to resist invasion by alien species more successfully—an important advantage in today's world of invasive species. For example, forests that were mature and undisturbed before most of the invasive alien plants escaped cultivation in North America still remain relatively free of alien plants because it has been difficult for these species to find space not already dominated by a diverse array of natives.

Are introduced species good or bad?

You might wonder why, if diversity is such a good thing, everyone is so concerned about the addition of species that evolved elsewhere to North American ecosystems. This is a good question, which scientists are still debating. To date, some 50,000 alien species of plants and animals have colonized North America. Not all are invasive—or not yet—but many have penetrated every natural area in the country. Some people claim that alien species actually increase the diversity of the region they have invaded and therefore are ecologically desirable. I could not disagree more.

If you simply count numbers of species, the species diversity of an area does rise after aliens invade, at least until they outcompete the natives for space, light, water, and nutrients. Diversity, however, is not a panacea in and of itself. The benefits of diversity enumerated above are not realized unless the species in an area are functioning members of an interacting community. If an alien plant penetrates a native plant community without decreasing the density of any of the existing species, its presence in that community may well be benign. Often, however, aliens colonize disturbed areas faster than do native plants and then subsequently prevent native colonists from reestablishing the plant community that was in place before the disturbance. For an alien species to contribute to the ecosystem it has invaded, it must interact with the other species in that ecosystem in the same ways that the species it has displaced interacted. In other words, it must be the ecological equivalent of the native species it replaces.

Let's take as an example the invasion of the Florida Everglades by *Melaleuca quinquenervia*, the paperbark tea tree. *Melaleuca* was imported from

Australia as an ornamental tree in 1906. It is a highly adaptable genus and can grow well in either dry soil or standing water. When humans lowered the water table over much of the Everglades, *Melaleuca* quickly invaded, and it has now almost completely displaced native grasses on hundreds of thousands of acres (Gordon 1998). Where *Melaleuca* has invaded, our treasured "River of Grass" has become a dense forest of alien trees that are completely useless to the creatures of the Everglades.

When I say *Melaleuca* has almost completely displaced native vegetation where it has invaded the Everglades, the operative word is "almost." If you were to measure plant diversity in a *Melaleuca* grove, you would probably find a few isolated individuals of most of the plant species that dominated the area before its arrival. If you were particularly hard-nosed, you might look at the entire Everglades and conclude that *Melaleuca* has not caused the extinction of a single plant species. Local extinction doesn't matter if one is simply listing the number of plant species that exist in an entire region. Your survey might actually show that plant species diversity has increased by one since the arrival of *Melaleuca* in the Everglades.

What would be missing from your survey, though, is a measure of what *Melaleuca* is contributing, or more aptly, failing to contribute to the Everglades ecosystem, compared to the grasses it has displaced. *Melaleuca* has transformed the sunny wet grasslands it has invaded into deeply shaded, drier forests dominated by a single species. The grassland birds that breed in the Everglades cannot nest in *Melaleuca* groves, and they find fewer insects to eat because native insects cannot eat *Melaleuca* leaves (Costello et al. 1995). Alligators cannot make their wallows or find food in *Melaleuca* groves, and so they have lost hundreds of thousands of acres of habitat. Butterflies cannot find their host plants, egrets cannot hunt the fish they eat, and hummingbirds cannot find the nectar they need to survive from day to day. Even though the number of plant species present in the area is higher by one, the ecological interactions that drive the Everglades ecosystem have collapsed where *Melaleuca* has invaded because its dominance contributes nothing to other organisms. Diversity is important, then, only when the species that can be said to create it are contributing members of their ecosystem. This contribution is most likely when species have evolved together over long periods of time.

The example of *Melaleuca* underscores how important it is to keep aliens that do not function within ecosystems from displacing the plants and animals that do have critical roles in their ecosystems. This is best done by maintaining the full diversity of native organisms in an ecosystem. It is well known that alien organisms invade faster and do better in "disturbed" sites. Translate "disturbed" to mean areas from which species, whether a few or many, have been removed. Although natural disasters happen, the disturbances I am talking about are almost always wrought by humans.

As we have seen, scientific analyses of ecosystems and the communities within them support the hypothesis that ecosystems work better when they are more diverse. Studies by dozens of scientists all suggest that ecosystems with more species function with more efficiency, are better able to withstand disturbances, are more productive, and can repel alien invasions better than ecosystems with fewer species (e.g., Elton 1927; MacArthur & Wilson 1967; May 1973; Paine 1980; Tilman, Wedin & Knops 1996; Wilson 1999; Tilman 2000; Moore et al. 2004; de Ruiter et al. 2005). But what about the people who live in New York City, or any other city for that matter? They seem to be doing all right, yet the places in which they live can hardly be described as diverse or healthy ecosystems. How does the Manhattan ecosystem function when the primary residents are cockroaches, house sparrows, pigeons, starlings, and ailanthus—all species from other lands?

Ecological sources and sinks

This question forces us to look at the concepts of ecological sources and sinks. If New York City were an isolated entity without connections to other parts of the country, it *would* collapse—in less than a week. Manhattan Island is an ecological sink; it requires the influx of great quantities of ecological resources that are generated in healthy ecosystems elsewhere (ecological sources) to sustain life. Manhattan does not trade natural resources; it only takes them. Manhattan Island does not have enough of its own water or food to support more than a few thousand people, and it certainly does not produce the surplus of ecological resources needed to export ecosystem services. It is not, by itself, a sustainable ecosystem. Too much of the biosphere covering Manhattan—that thin zone in which conditions are just right for life—has been destroyed by blacktop, exhaust, and skyscrapers. People can live in New York City only because they take what they need to live from areas of the country that still have a healthy biosphere. The water that quenches the thirst of millions of New Yorkers comes entirely from an ecosystem that remains functional: the forested Catskill Mountains north of the city. The oxygen that New Yorkers breathe is generated by the vast forests of the Amazon and by populations of phytoplankton in the sea. The fish that are served in exclusive West Side restaurants come from oceans all over the world. The beef comes from rangelands across the continent, and the grain from fragile topsoils laid down during the last glaciation in the Midwest. Every natural resource required to keep New Yorkers alive comes from ecosystems that have not yet collapsed. If urban and suburban sprawl destroys the hydrology of the Catskills, New Yorkers will suffer. If we continue to employ the farming methods that lose tons of topsoil per acre every year, New Yorkers will be breadless. If we continue to overharvest the world's great fisheries, New Yorkers will lose fish as a source of protein. If we convert the rest of the world to an artificial habitat fit for humans but nothing else, New Yorkers, as well as the rest of us, are doomed.

I wish these were preposterous scenarios, but we are enacting them as I write. When I talk about the value of biodiversity in suburbia, I am talking about a natural resource that is critical to our long-term persistence in North America. Biodiversity is a national treasure that we have abused terribly, partly because we have not understood the consequences of doing so. Our understanding of such consequences is far from perfect, but we now know enough to behave responsibly toward the plants and animals on which we ourselves depend. We must manage our biodiversity just as we manage our water resources, our clean air, and our energy. Fortunately, unlike most of our water or energy supplies, biodiversity is a renewable resource that is relatively easy to increase, as long as we do so before its components, the species themselves, become extinct.

CHAPTER FIVE

Why Can't Insects Eat Alien Plants?

When I was a boy, driving at night during the summer would invariably produce a blizzard of nocturnal insects in the car's headlights. Today I see only the occasional moth flutter by. This apparent decline in insect populations is being noticed all over the temperate zone and has been scientifically confirmed in Great Britain (Conrad et al. 2006). In the past I have attributed the loss of insects to habitat loss, and surely the loss of natural areas has played an important role in the reduction of insect populations. Then I noticed that insects seemed to prefer native plants over alien species, and another reason for their decline in our gardens and backyards was suddenly obvious.

A black walnut tree covered with multiflora rose and Japanese honeysuckle. The aliens block sunlight from the walnut leaves in the summer and weigh the branches down during ice storms in the winter. Trees engulfed to this degree cannot perform their role in the ecosystem, and they are eventually killed by the aliens.

"Pest-free" ornamentals are favored

But why, you ask, can't insects use alien plants for growth and reproduction? Ecologists suggest three reasons why most native insects cannot, or will not, eat alien plants. First, many of the alien plants that have succeeded in North America are not a random sample of all plants that evolved elsewhere, but rather are a subset that were imported specifically because of their unpalatability to insects (Tallamy 2004). As Michael Dirr repeatedly emphasizes in his acclaimed volume on ornamental plants (1998), species that are "pest free" are favored by the ornamental industry. What no one counted on when these pest-free ornamentals were brought into the country was that many would subsequently escape cultivation and replace native plants in natural areas everywhere. In the South and East, some of the worst offenders in this category include several honeysuckle species, the *Melaleuca* paperbark tree, autumn olive, privet, multiflora rose, kudzu, lantana, buckthorn, oriental bittersweet, purple loosestrife, Norway maple, burning bush, English ivy, Japanese knotweed, Bradford pear, empress tree, Japanese barberry, wisteria, and mile-a-minute weed. These and many more invasive plants were introduced to this country by well-meaning horticulturists looking for exciting new species to sell in the garden trade (Mack & Erneberg 2002).

The slow pace of adaptation

The second reason native insects do not use alien plants should come as no surprise: it takes time—long evolutionary time spans rather than short ecological periods—for insects to adapt to the specific chemical mix that characterizes different plants (Kennedy & Southwood 1984). When plants arrive at our shores, they bring with them few or none of the insects and pathogens that curb their growth at home. As Mark Williamson explains in his book on biological invasions (1996), they have been "released" from the constraints of their natural enemies and can therefore outcompete most native plants that support a full complement of herbivores and pathogens. These unnatural introductions create perfect opportunities for measuring how long it takes native insects to adopt alien plants as hosts. The literature on plant-insect interactions is replete with evidence that the number of herbivores associated with transplanted aliens is only a small fraction of the historical complex of natural enemies that kept these plants in check at home. In Europe, for example, the Eurasian genotype of *Phragmites australis* supports

Phragmites australis supports 34 times more insect species in its European homeland than it does as an invasive form in North America.

over 170 species of phytophagous insects, while only 5 species of our native herbivores use this plant in North America (Tewksbury et al. 2002). Since its introduction to Florida from Australia in the early 1900s, only 8 species of arthropods have been recorded eating *Melaleuca quinquenervia* leaves. In Australia, however, where this species is actually rare, 409 arthropods are known to eat *Melaleuca* (Costello et al. 1995). *Eucalyptus stellulata* is host to 48 insect species in Australia (Morrow & LaMarche 1978), but only 1 in California (Strong, Lawton & Southwood 1984). Similarly, *Clematis vitalba* supports 40 species of insect specialists in its European homelands, but only a single native insect has adopted it as a host in New Zealand (Macfarlane & van den Ende 1995). Even after several centuries of exposure, no insects eat the cactus *Opuntia ficus-indica* in South Africa. Sixteen species specialize on this plant in the Americas (Annecke & Moran 1978). Five hundred years after conifers in the families Cupressaceae, Pinaceae, and Taxaceae were introduced into Great Britain, only 50 of the 500 species of moths available to eat these species actually do so (Winter 1974). These examples clearly demonstrate that adaptation by our native insect fauna to plant species that evolved elsewhere is a slow process indeed.

Most insects are specialists

The third reason I was not surprised to see native insects shunning aliens is that scientists who know what they are talking about say they should do just that. Since the 1960s, an extensive body of theory developed by experts on interactions between plants and insects predicts that most phytophagous insect species should be able to eat only vegetation from plants with which they share an evolutionary history (Ehrlich & Raven 1965; Rosenthal & Janzen 1979; Strong, Lawton & Southwood 1984; Berenbaum 1990). Up to 90 percent of all phytophagous insects are considered specialists because they have evolved in concert with no more than a few plant lineages (Bernays & Graham 1988). The chrysomelid beetle genus *Phyllobrotica* provides a perfect example of specialization (Farrell & Mitter 1990). Members of this genus eat plants only in the genus *Scutellaria*, the skullcaps. What's more, each species of *Phyllobrotica* eats only one species of *Scutellaria*. If skullcaps disappear from an area, so do their *Phyllobrotica* specialists.

I have encountered in my own yard an excellent demonstration of just how restricted a specialist's diet is. A young black cherry tree had been completely defoliated by a group of tent caterpillars whose mother had unwisely chosen to lay her eggs on a tree too small to support her offspring for their entire development. When the cherry had been stripped of its leaves, the caterpillars were forced to leave the tree in search of another suitable host plant. This undoubtedly was a very dangerous undertaking for the caterpillars, not only because it exposed them to predators, but because the likelihood of actually finding another host before they starved to death was slim. I mention the risks involved in abandoning a host plant to emphasize that the caterpillars would not have left the tree if there had been any edible leaves left for them. What is interesting in this case is that there were still leaves available on the cherry. Winding up the trunk and out along the branches was a large Japanese honeysuckle with plenty of foliage. The caterpillars must have walked over the honeysuckle leaves repeatedly to find every last cherry leaf, and yet they had not taken a single bite of the alien plant. Even as they ran out of food, the caterpillars simply did not recognize honeysuckle

Beetles in the genus *Phyllobrotica* only eat plants in the genus *Scutellaria*.

as a potential food source. Eastern tent caterpillars (*Malacosoma americanum*) are specialists on a single lineage of plants in the order Rosales; plants outside that lineage cannot serve as food for these insects, even in the face of starvation.

Such restricted host plant associations typically require long periods of evolutionary time to develop (Kennedy & Southwood 1984). To exploit a plant effectively, insects must evolve the ability to find their particular host species amid thousands of other plants and then to synchronize their life cycle with the appearance of the needed parts of their hosts. Insects must also develop the ability to overcome the physical and chemical defenses of their hosts through behavioral and physiological adaptations. The evolution of specialized abilities to eat the tissues of one particular plant lineage usually decreases an insect's ability to eat other plants that differ in timing of development, leaf chemistry, or physical defenses such as trichomes or tough leaves. By definition, native insects have shared little or no evolutionary history with alien plants (although some may have interacted with a species in a common genus), and they thus are not likely to possess the adaptations required for using these plants as nutritional hosts (Tallamy 2004). Consequently, the solar energy harnessed by alien plants is believed to be largely unavailable to native insect specialists—at least until they evolve the behaviors and physiology necessary to eat them—and therefore unavailable to all animals that include these insects in their diets.

Exceptions to the rule?

Does this mean that insect specialists are never able to eat and reproduce on alien plants? No, there are specialists that have "adopted" an alien plant species as a host after a relatively short period of exposure to the plant. One good example is the broad-winged skipper (*Poanes viator*), a native lepidopteran (butterflies and moths) that now includes the European genotype of *Phragmites australis* as a host plant. Another good example is seen in the use of dill and parsley by the black swallowtail (*Papilio polyxenes*). Dill and parsley are not native to North America, and yet the black swallowtail develops quite nicely on them.

In both of these examples, what initially appears to be an anomaly turns out to be exactly what theory predicts. The broad-winged skipper and the black swallowtail have been able to use *Phragmites*, and dill and parsley, respectively, even though they have shared no direct evolutionary history with these plants because they have all of the adaptations necessary to find and digest them. That they possess the required adaptations to use these aliens as hosts is not just a happy coincidence for the broad-winged skipper and the black swallowtail. Instead, it is the product of millions of generations of specialization on native plants that are close relatives of *Phragmites*, dill, and parsley. The broad-winged skipper is a grass specialist and may even have eaten the native *Phragmites* genotype before it was displaced by the alien genotype. It is able to recognize and use European *Phragmites* because *Phragmites* is a grass, just like its native hosts. Likewise, the black swallowtail is a specialist on North American representatives of the carrot family (Apiaceae). Dill and parsley are also members of the carrot family. Even though they did not evolve in North America, parsley and dill contain the complex mix of chemicals in their leaves that characterize all members of the Apiaceae and allow the black swallowtail to smell these plants, eat them, and digest their tissues.

It is by no means guaranteed that insect specialists from North America can use all aliens that are related to their native host plants, but almost all cases in which aliens have been adopted by native specialists involve plants closely related to the insect's native host species. One of the most active areas of research today is to learn to what extent such host adoptions are

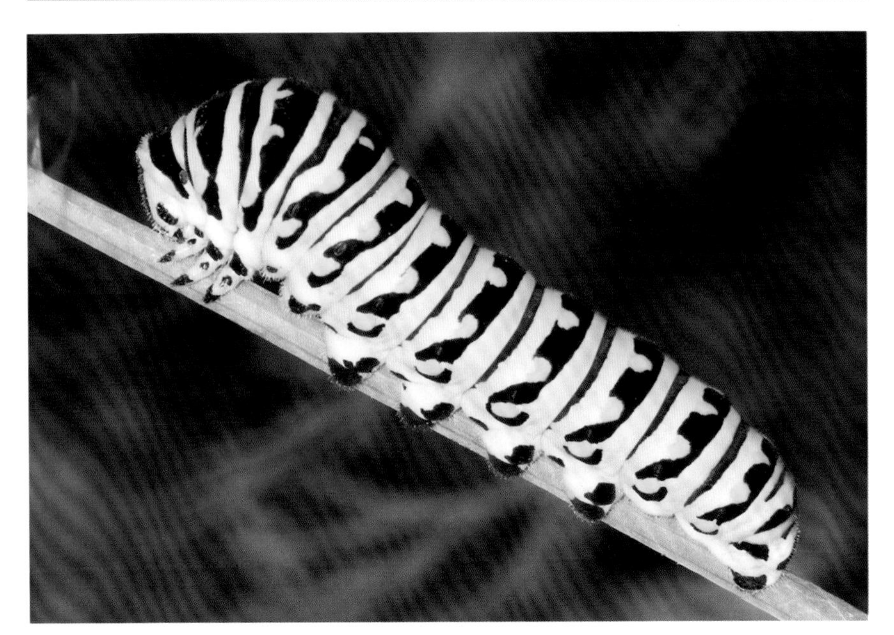

The larva of the native black swallowtail (*Papilio polyxenes*) can develop on alien dill because dill is a close relative of the swallowtail's native hosts.

happening across the country. Unfortunately, many alien plants are not closely related to any plant species in North America. Plants such as English ivy (*Hedera helix*), goldenrain tree (*Koelreuteria paniculata*), crape myrtle (*Lagerstroemia indica*), European privet (*Ligustrum vulgare*), katsura tree (*Cercidiphyllum japonicum*), empress tree (*Paulownia tomentosa*), chocolate vine (*Akebia quinata*), and dozens more species in the ornamental trade are not related to any lineage of plants in North America. It is highly unlikely that insect specialists will be able to use such plants any time soon.

Why leaf chemistry drives insect specialization

Of the many obstacles that prevent insect specialists from finding, eating, growing, and reproducing on alien plants, leaf chemistry is probably the hardest to overcome. The chemical constituents that allow plants to fix energy from the sun and turn that energy into new tissues, a process called "primary metabolism," are nearly identical across the plant kingdom. Yet if you were to grind up a leaf from each of the species of plants on earth and analyze the array of chemical compounds present in each leaf, you would find that no two species have the same leaf chemistry. There are many compounds in each leaf that have nothing to do with primary metabolism. These so-called secondary metabolites differ in each species, either in the structure of the compounds themselves or in the combinations and amounts of the compounds present. Secondary metabolites give each plant species a particular and unique taste, digestibility, and toxicity.

COMMON SECONDARY METABOLIC COMPOUNDS FOUND IN THE LEAVES OF PLANTS

Chemical Class	Plant Sources
Glycosides	
cyanide glycosides	almonds, cassava, lima beans
iridoid glycosides	*Plantago, Lagochilus, Incarvillea*
cardenolides	milkweeds, *Isoplexis, Digitalis*
glucosinolates	broccoli, cauliflower, rapeseed
Phenols	
coumarins	Tonka bean, lavender, licorice
tannins	oaks, beech, hickory
lignins	grasses
Terpenes	
cucurbitacins	cucurbits, candytuft
limonoids	neem, *Carapa*
saponins	yucca, daisies, horse chestnut
Alkaloids	
benzylisoquinoline	poppy, *Colchicum*
pyrrolizidine	composites, legumes
quinolizidine	*Lupinus, Nicotiana, Conium*
nicotine	tobacco, eggplant, tomato

Several hypotheses attempt to explain why plants make chemicals that they don't need to bloom and grow, but most plant ecologists agree that an important function is defense against herbivores (Rosenthal & Janzen 1979). In essence, "herbivores" means insects, since they are the primary plant-eaters in most parts of the world (Wilson 1987). Most secondary compounds are

The caterpillar of the monarch butterfly (*Danaus plexippus*) is a specialist on members of the milkweed family (Asclepiadaceae). Without milkweeds in the landscape, there will be no monarchs.

nasty things that make a leaf distasteful at the least, and typically toxic to all animals that have not developed the enzymes needed to detoxify them. Insects that do have specialized adaptations for detoxifying a particular plant's leaf chemistry, usually a few dozen insect species, are not deterred; but many thousands of insect species do not possess those particular adaptations and cannot exploit that plant at all.

Our own ability to eat plants provides a good analogy. Dozens of common plants would make us quite sick if we were to eat them (Kingsbury 1964). Black cherry leaves, pokeweed leaves, and deadly nightshade all fall into the "poisonous" category for people. The tannins in oak leaves would bind up all of the protein we ingested, eventually starving us to death. Even many of the crops we do eat originally had toxic chemicals in their leaves before we removed them through plant-breeding programs that have been in progress for thousands of years. When we first domesticated lima beans, we discovered the hard way that if we don't boil the beans, we risk a lethal dose of cyanide. The cucurbitacin triterpenes in cucumber leaves would kill us in short order if we could choke them down. And we all know the consequences of consuming nicotine from tobacco or opiates from poppies. In most cases, though, we have no trouble avoiding leaves containing such compounds because they taste so bad. Our lettuce has to be young and fresh or it gets too bitter for our fussy taste buds. We have all eaten cucurbitacins in minute quantities in the occasional bitter cucumber, but we don't make that mistake twice in a row.

My point here is that insects suffer the same constraints. They can only tolerate a narrow group of chemicals to which they have been repeatedly exposed over thousands of generations. All other chemicals taste bad to insect herbivores and signal that the insect is on the wrong host. Specialists do not have the option of eating any plant in the neighborhood. Nor do they have the option of quickly adapting to the alien plants that have replaced their native hosts. Evolution simply does not work that fast. If an insect's hosts are not present, it won't be either.

Can generalists compensate for the loss of specialists?

So far I have described well-accepted hypotheses about how evolution has molded most insect herbivores to become "specialists": insects that are adapted to find, eat, digest, and survive on plant lineages that produce particular types of phytochemicals. I have said little, though, about the species that are not specialists on particular host plants, about 10 percent of the insect herbivores in a given ecosystem. In contrast to specialists, "generalist" insects have evolved the ability to eat several types of plants. Toxicologists have learned that one of the adaptations that permit some insects to generalize on many host plants is the ability to produce very powerful gut enzymes called mixed-function oxidases (Brattsten, Wilkinson & Eisner 1977). These enzymes are so good at detoxifying different classes of plant-defensive chemicals that generalists can eat lots of unrelated plants with no measurable reduction in their ability to grow and reproduce.

If 90 percent of the species of insect herbivores are specialists, why are the 10 percent that are generalists important when considering the impact of alien plants on the food web? Generalists may play an important role in mitigating the effects of alien plants for two reasons. First, although there are lots of species of specialists, most are relatively uncommon. Second, perhaps because they have the luxury of eating many species of plants, the few species of generalists out there are usually quite common (Futuyma & Gould 1979) and are thought to have a disproportionately large impact on plant communities (Crawley 1989; Hay & Steinberg 1992). In combination, these two facts mean that the amount of biomass that insect specialists contribute to the food web at any one time may be relatively small compared to the contributions of insect generalists. What's more, there is good evidence that vertebrate generalist herbivores (Parker & Hay 2005) and maybe even some invertebrate generalists (Parker, Burkepile & Hay 2006) actually prefer alien plants as food, possibly because alien plants have not evolved effective defenses against novel generalists in the lands they invade (Hokkanen & Pimentel 1989). Therefore, if generalist insect herbivores do not discriminate against alien plants when selecting their next meal (or actually prefer them), it is theoretically possible that alien plants could replace natives in a given habitat without seriously depleting the insect biomass available to higher trophic levels. There would be fewer insect species in this hypothetical environment, but more of them would be common generalists and the number of insect bodies for birds, bats, and other insectivores to eat might be about the same as the number found in a habitat comprised entirely of native plants. If, on the other hand, the leaf chemistry of alien plants is too foreign and thus too unpalatable even for generalists, both specialists and generalists would decrease in abundance in a habitat invaded by alien plants and the insect biomass available for higher trophic levels would be seriously depleted.

How do we know the actual extent to which our native insect generalists are eating alien plants? We don't until we go into the field and see exactly what is eating what. Unfortunately, this important but simple task has been all but ignored so far. My students and I have been working to fill this gap in our knowledge, and despite the fact that most of our studies have not yet been completed (they take years!), preliminary data are starting to accumulate (for details of my analyses, see appendix 3). One of the first things I did was compare the diversity and biomass of the insects that were developing on the four most common woody natives in our yard (black oak, black cherry, black walnut, and fox grape) with the insect diversity and biomass on our five most common alien plants (autumn olive, mile-a-minute weed, oriental bittersweet, multiflora rose, and Japanese honeysuckle). Following standard protocol for sampling diversity, I found that native plants produced over 4 times more herbivore biomass than did alien species and supported 3.2 times as many herbivore species. When I compared natives and aliens in terms of their production of Lepidoptera (moths and butterflies) and sawfly caterpillars—the largest diet component of insectivorous birds—I found that the

The bitter cucurbitacins found in the tissues of cucurbits are toxic to most animals, but the western striped cucumber beetle (*Acalymma trivittata*), a specialist that only eats cucurbits, is one of many beetles that have evolved the ability to detoxify these compounds.

Although most insects are physiologically capable of eating only a few types of plants, about 10 percent can eat the leaves of several unrelated plant groups. Generalists such as the white-marked tussock moth (*Orgyia leucostigma*) may supply a large percentage of the insect biomass eaten by animals in higher trophic levels.

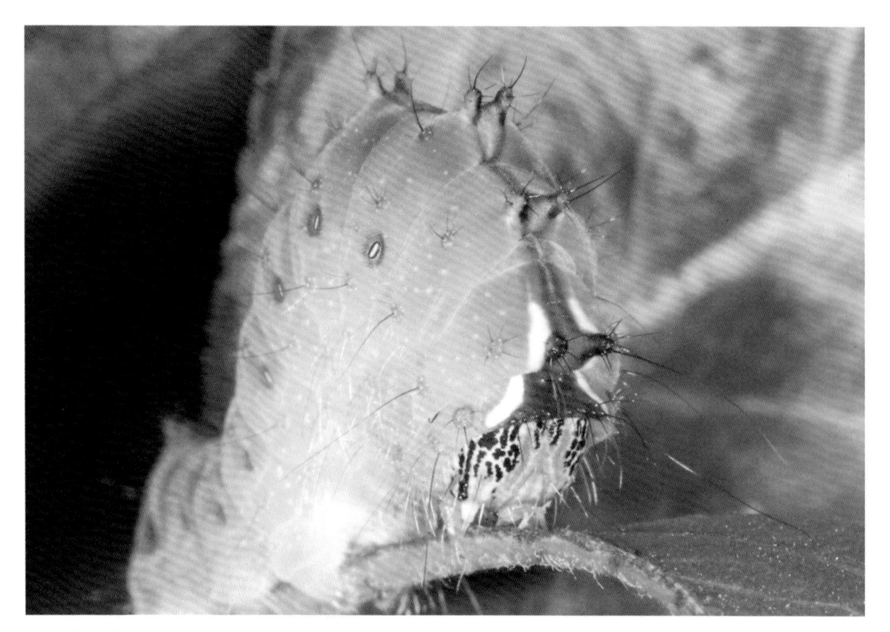

The clear dagger moth (*Acronicta clarescens*) is a generalist capable of eating many species of the rose family (Rosaceae), but it cannot digest tree of heaven, autumn olive, barberry, Japanese honeysuckle, oriental bittersweet, burning bush, privet, or any other pervasive alien.

native plants in the study supported a whopping 35 times more caterpillar biomass than the aliens. We know that most bird populations are limited by the amount of food they can find (Marra, Hobson & Holmes 1998; Duguay, Wood & Miller 2000; Nott et al. 2002), so if there is 35 times less food available for birds in habitats that comprise primarily alien plants, there will be 35 times less bird biomass in those habitats as well. No wonder our birds are struggling.

Data from this experiment can also be used to determine whether generalist insects are compensating for the loss of specialists in the production of insect biomass. Because I had collected all of the insects on the target plant species, I could identify them to species and compare the biomass of generalists produced on aliens with that produced on native plants. After eliminating all of the specialists produced on natives (as predicted, the alien plants in my study produced no specialists), I found that natives still produced more than twice as much generalist biomass as did aliens. That is, I found no evidence that generalists are able to use alien plants at the same rate they use natives, and I certainly did not find that generalists *prefer* to eat alien plants.

Although luna moth larvae are by definition "generalists," my analyses show that they are not physiologically capable of using either alien plant species or most of the plants recorded as bona fide native host plants for growth and development. Lunas from any given population behave just like specialists.

The role of generalists in providing insect biomass for birds and other creatures in habitats invaded by alien plants is such an important issue that it must be examined with several well-designed experiments before we can feel comfortable that we truly understand it. My students and I are continuing to look at the question, although our first experiments suggest that generalists will not be able to compensate for the loss of specialist species in invaded habitats. In her 2006 master's thesis for the University of Delaware, Meg Ballard addressed the question of how generalists respond to aliens in field habitats. In a two-year common garden experiment, she compared the biomass of insects produced on six herbaceous aliens (lambsquarters, cocklebur, velvetleaf, jimsonweed, pigweed, and cosmos) with the insect biomass on six herbaceous natives (eastern black nightshade, black-eyed Susan, devil's beggarticks, ragweed, horseweed, and goldenrod).

Out of 93 species of insect herbivores collected on these plants, Meg identified only two species as being specialists (Ballard 2006). This was not surprising because early successional habitats—habitats that are being recolonized by plants after a disturbance—are exploited more by generalists than by specialists (Strong, Lawton & Southwood 1984). We reasoned that if generalist insects are using alien plants as much or more than native plants anywhere, it would be in field habitats, and that Meg's study would document it. Instead, the data convincingly showed the opposite pattern of herbivory (see appendix 3). Native plants produced nearly six times more generalist insect herbivore biomass than did alien plants, particularly as population sizes built over the summer. There was no evidence that generalists prefer aliens, nor that alien plants are capable of producing as much insect biomass (bird food) as native plants.

Do alien plants harm birds?

For many of us, the most compelling reason to return native plants to the landscape is the role such plants have in producing food for our charismatic fauna, particularly our birds. But is there any hard evidence that birds suffer when alien plants replace natives? As important as this question is, there are few studies that have addressed it. No doubt the difficulty of this type of research has much to do with the lack of data. Measuring reproductive success in birds is challenging under the best circumstances, but being able to attribute reproductive success to a particular environmental factor is even more difficult. In addition to the quality and quantity of food in the environment, there are dozens of things that influence how successful a bird might be in producing and rearing offspring. These include predation levels, bird age and experience, weather, and the level of competition from birds in the same and different species.

Nevertheless, two studies have been published that document bird declines in areas heavily invaded by alien plants. In the first study, Aron Flanders and his coworkers compared bird communities in replicated 200-hectare plots (about 494 acres) within two South Texas rangeland habitats (Flanders et al. 2006). The first habitat was made up almost entirely of native grasses and forbs, whereas the second was dominated by two invasive alien grasses: Lehmann lovegrass (*Eragrostis lehmanniana*) and buffelgrass (*Cenchrus ciliaris*). Both of these species were introduced in the 1940s to "restore" overgrazed rangeland. Their introduction was a "success," and today they have replaced native vegetation in millions of acres throughout the southwest (Ibarra-F. et al. 1995). The Flanders study showed that restructuring the grassland plant community with alien plants has, in turn, restructured the insect community and the birds that depend on insects at the study sites. Both arthropods

Mockingbirds (*Mimus polyglottos*) have been found to be twice as abundant in native prairies than in prairies invaded by buffelgrass and lovegrass (Flanders et al. 2006).

(insects and spiders) and insectivorous birds were significantly more abundant on the plots that still had native vegetation (60 percent and 32 percent respectively) as compared to plots invaded by lovegrass and buffelgrass.

A second study, conducted at Medicine Lake National Wildlife Refuge in Montana, suggests that the results of the Flanders study do not represent an exceptional pattern but will more likely prove to be the rule. John Lloyd and Thomas Martin (2005) compared the breeding success of the chestnut-collared longspur (*Calcarius ornatus*) in patches of native prairie and patches of prairie invaded by crested wheatgrass (*Agropyron cristatum*), a species introduced from Asia that as of 1996 had spread over 10 million acres in the West (Lesica & DeLuca 1996). They found that longspur nestlings reared in areas dominated by wheatgrass grew more slowly and were smaller than nestlings reared in native prairie vegetation, an obvious sign that food was in short supply in alien wheatgrass. Being small is dangerous in nature, and nestlings in the wheatgrass were 17 percent more likely to die on any given day than nestlings in native prairie grasses. The results of these two studies should come as no surprise: removing the food (native plants) of the food (insects) that birds need to rear their young will result in fewer birds.

CHAPTER SIX

What Is Native and What Is Not?

When I first became interested in the impact of alien plants on our nation's biodiversity, I was surprised to learn that the terms "native" and "alien" were controversial. Somewhat naively, I thought that when a plant found in China is sold in the United States as an ornamental, it can be classified as an "alien" without much debate. I was wrong. Recently a good friend insisted that if a plant has been in North America long enough, it can be considered a native, regardless of its evolutionary origins. Even more problematic is achieving consensus on the definition of a "native" plant. The broadest definition is also the most commonly employed: a native is any plant that historically grew in North America. Some people recognize that it's a stretch to call a plant adapted to a California desert a native of New Jersey, but these same people happily consider all plants that grow east of the Mississippi as native to any area in the East. Maps of U.S. hardiness zones are often brought out to justify the "nativeness" of one species or another. By this reasoning, a plant adapted to zone 6 in Tennessee will serve as a fine native in the areas of Pennsylvania with a zone 6 climate.

Let nature define nativity

The problem with these definitions of "alien "and "native" is that they do not consider the roles plants play within their respective ecosystems. I believe that what is and is not a native plant is best defined by nature herself. Because plants do not grow in isolation from the other living things around them and are in fact essential to the lives of neighboring creatures, they interact with the residents of their habitats in countless ways. Over immense periods of time, these interactions help shape both the plants and the animals in a particular place. That is, the plants and animals in an area "coevolve," each group continually influencing the evolution of the other. When a plant is transported to an area of the world that contains plants, animals, and diseases with which it has never before interacted, the coevolutionary constraints that kept it in check at home are gone, as are the ecological links that made that plant a contributing member of its ecosystem. In an ecological and evolutionary sense, the alien plant's new neighbors won't know what to make of it and, in most cases, will exclude it from their biological interactions. The plant will occupy space and use resources (light, water, and soil nutrients) that would otherwise have been available for a native plant, but it will not pass the energy it harnesses from the sun up the food chain. If alien species were providing as many ecosystem services in their new homes as they did where they evolved, they would support about the same number of insect species in both areas. But as we have seen, this is hardly the case. This is why I argue that a plant can only function as a true "native" while it is interacting with the community that historically helped shape it.

When "native" and "alien" are defined in terms of the presence or absence of historical evolutionary relationships, the confusion over these concepts disappears. The Norway maple (*Acer platanoides*), for example, evolved in northern Europe. When John Bartram introduced the tree to the Morris Arboretum of Philadelphia in 1756, this species had never before interacted with the plants, animals, and pathogens of North America (Nowak & Rowan 1990). Well into its third century of residence, the growth and reproduction of Norway maples is still influenced little or not at all by organisms native to this country. Nor have its centuries here been long enough for North American plants and animals to adapt to the presence of Norway maple. Why has its residence made so little difference in shaping Norway maple to the needs of the plants and animals that evolved in North America? Quite simply, a history measured in centuries is the tiniest drop in the proverbial bucket of

Norway maple (*Acer platanoides*), an alien introduced into the ornamental trade from Europe, is now the most common shade tree in North America. As with many ornamental species, it has escaped cultivation and is rapidly displacing native trees.

evolutionary time. The lineage that gave rise to Norway maples in Europe has been separated from the lineage that gave rise to North American maples ever since the North American continent separated from Eurasia over 80 million years ago. During that immense span of time, Norway maples diverged genetically from North American maples in countless ways, including leaf chemistry, the timing of flower and seed set, growth rate, shade tolerance, and so on. Two or three hundred years is only the smallest fraction of the 80 million years Norway maple has been changing from its North American relatives. We can hardly expect Norway maple to fit productively into North American ecosystems after spending only a miniscule amount of its evolutionary history on this continent.

If we can agree that "native" and "alien" are not arbitrary terms open to debate, but instead are unambiguously defined by a plant's evolutionary background, it is clear that Norway maples and any other plant from Europe or China can only be considered as aliens. But what about plants from North America? Blue spruce and Douglas firs are good examples to consider. Blue spruce and one genotype of Douglas fir spent much of their evolutionary history in the Rocky Mountains; the other genotype of Douglas fir is found in the Pacific Northwest. When a resident of Portland, Oregon, plants a Douglas fir, the tree will grow within the community that historically helped to create it. But if we plant a Douglas fir in Oxford, Pennsylvania, the tree is thousands of miles from the community of organisms that were part of its evolutionary history. It will grow just fine in Oxford, but it will not function as a native to that area, regardless of any attempts to label it as such. A similar story can be told for blue spruce. We hear time and again from well-meaning homeowners and nurserymen that it is a wonderful native, albeit a little too blue for some garden designs. Indeed, it *is* a wonderful native—in western Colorado. But in the suburbs of the East it is as divorced from its historical ecosystem as China's "tree of heaven."

So far I have painted a pretty simplistic picture, so let's complicate things a bit. There are, in fact, cases where a plant can be moved outside of its native range and still perform some or even most of its evolutionary roles within its new ecosystem. This typically happens when a plant is a member of a genus that contains several similar relatives. Animals adapted to using one member of the genus are often able to use a close congener (a member of the same genus), even if they have never interacted with that particular plant species in their evolutionary past. This can happen because traits such as leaf chemistry, shape, and toughness can be so similar among congeners that adaptations enabling an insect to grow and reproduce on one member of the genus predispose that insect to using other members of the genus. Sundrops (*Oenothera fruticosa*), for example, are restricted to the Upper Piedmont and

Douglas fir (*Pseudotsuga menziesii*) evolved in the Pacific Northwest of North America, where it performs important ecosystem functions. Outside that regional habitat, however, it does not support diverse insect fauna and therefore does not supply as much food for other animals as it does in its native ecosystem.

the mountains of the Northeast. Even though they do not naturally occur in the Lower Piedmont habitats near Oxford, Pennsylvania, we can enjoy them on our property guilt-free, knowing that they will serve as suitable host plants for the many insects that depend on their more widespread relative, evening primrose (*O. biennis*), which does grow commonly in our area. The very showy azaleas that evolved in and around the Great Smoky Mountains in Tennessee and southwestern Virginia are becoming popular ornamentals in many other areas. They can remain a functioning part of the ecosystem to which they are moved because insects adapted to local azalea species such as pink (or pinxter) azalea (*Rhododendron periclymenoides*) or swamp azalea (*R. viscosum*) should have no trouble using the southern species as a resource. My point here is that a gardener need not be a complete purist in the use of native plants in recreating functioning habitats for insects and the many birds and animals that eat them.

Can alien plants that are related to natives support our insects?

You might now wonder, "If I can plant an azalea from Tennessee in New Jersey without a serious loss to local biodiversity, why can't I plant an azalea from China without creating problems?" The answer to this logical question is that maybe you can, but the chances of an azalea from China being so similar to an azalea from New Jersey that New Jersey insects will be able to adopt it as a host plant are pretty slim, even if the plants are in the same genus. Any two plant species that have been isolated from each other on different continents for many millions of years are likely to have developed leaf chemistries different enough that either the insects searching for a host plant won't be able to recognize the alien as a suitable host, or if they do, they won't be able to eat and digest the plant safely. These are not hard and fast rules. We suspect that research into these questions will identify a few alien plant species that are acceptable hosts to native insect specialists as well as to generalists. But these will be the exceptions and should not overly influence our general approach to alien species in our gardens.

Oenothera fruticosa, the lovely sundrops of the Appalachians (top), does not occur naturally in Delaware, but its range does overlap that of *O. biennis*, the evening primrose (bottom). Insects adapted to one species of *Oenothera* are therefore likely to be adapted to the other species, even if it is planted slightly outside of its natural range.

CHAPTER SEVEN

The Costs of Using Alien Ornamentals

So far I have focused on one detriment associated with the overuse of alien ornamentals in suburban gardens: the cost to wildlife as native food sources disappear. In my view, this is reason enough to increase the percentage of natives in our gardens. But there is another serious cost that arises when we garden with aliens that we can no longer ethically ignore. When gardeners support the market for plants from other countries, they encourage the introduction of alien stock to North America, with two serious consequences. First, despite our best efforts to bring only "clean" plants into this country, we continually introduce harmful diseases and insects that evolved on other continents along with our beloved alien ornamentals. The problems stemming from the introduction of an unwanted disease or insect pest are obvious. Less obvious (in fact, downright controversial) is the suggestion that gardening with alien ornamentals places our native ecosystems at risk of destruction by yet another invasive plant species.

Alien plants can be Typhoid Marys

Most of us know little about the historical track record of the ornamental industry in terms of bringing serious and sometimes disastrous pests to this country along with alien plant stock. There are dozens of examples of plant diseases that were inadvertently brought to our shores with nursery stock, but none has been more devastating to our eastern deciduous forests than the chestnut blight. The chestnut blight is of Asian origin and was transported to the Northeast in 1876 on *Castanea crenata*, resistant Japanese chestnut trees for ornamental trade (Powell 1900; Anagnostakis 1987). Lesions from the fungus *Cryphonectria parasitica* were first found on an American chestnut tree at the Bronx Zoo in 1904. Within 50 years, *Castanea dentata*, the dominant upland forest tree species from Maine to Mississippi—a tree that had previously survived an asteroid impact and at least 20 glaciations during its 87 million years of evolutionary history (Willis & McElwain 2002)—was functionally eliminated from the eastern deciduous forest ecosystem. One factor in the rapid spread of the fungus from Japanese chestnuts to our native chestnuts was the fledgling mail-order nursery business, which shipped thousands of infected Japanese trees up and down the East Coast to homeowners who were eager to try "Japanese Giant" chestnuts in their gardens (Graves 1930).

It is hard to overemphasize the impact that the loss of American chestnuts has had on deciduous forest ecosystems. *Castanea dentata* was the primary nut producer of eastern forests, dwarfing the contributions of oaks, beeches, and hickories as wildlife food sources. Squirrels, chipmunks, deer, elk, black bears, turkeys, passenger pigeons, doves, blue jays, and mice were just some of the animals that depended on copious quantities of chestnuts to make it through long winters. Equally important but more poorly documented was the role American chestnuts played in producing insects that supported huge populations of songbirds. *Castanea dentata* is a member of the plant family Fagaceae, which supports hundreds of species of caterpillars and other insects eaten by birds. The niche vacated by chestnut trees was slowly filled with other tree species (Cech 1986), and our forests appear today as though nothing had happened—unless you compare the density of wildlife that forests support today with what they used to support. Such a study is now impossible, not only because the chestnut is gone—along with an unknowable number of insect specialists that ate only the chestnut—but also because our forests are now so fragmented. Nevertheless, because none of the species that replaced the chestnut is as productive as the chestnut (Diamond et al. 2000), I predict that such a comparison would show smaller populations of both birds and mammals in today's eastern forests. For example, tulip tree (*Liriodendron tulipifera*) is one tree species that has done particularly well in the absence of chestnut, and it is now the dominant species in many forests of the East (Cech 1986). Unfortunately, it is one of the least productive forest species in terms of its ability to support wildlife—insects and vertebrates alike.

Some might argue that the chestnut blight was a one-time fluke that cannot happen again. After all, in response to the chestnut blight, the Plant Quarantine Act of 1912 was passed, specifically to prevent such disasters (Waterworth & White 1982). The microscopic nature of plant diseases, however, renders acts of Congress ineffective in keeping pathogens out of the country. Plant species that are imported for the ornamental trade often carry disease organisms to which they are resistant and show no symptoms. For example, dogwood anthracnose (*Discula destructive*) was first detected in the 1970s on flowering dogwood (*Cornus florida*). Although the fungus can be lethal to flowering dogwood, kousa dogwood (*C. kousa*) expresses no symptoms of the disease. It is highly likely that dogwood anthracnose was imported to the United States on kousa dogwoods from Asia because it escaped detection at quarantine facilities (Palm 2001). Like Typhoid Mary, ornamental kousa dogwoods may have spread a deadly disease with no one even knowing it. As long as we import hundreds of thousands of plants every year, there is real danger of introducing new diseases, perhaps even the next chestnut blight.

A recent contender is another fungus, first described on ornamental rhododendrons in Germany and then imported to the United States in nursery stock (Garbelotto 2004). *Phytophthora ramorum*, now called sudden oak death disease, was identified in the California nursery trade in 1995. It quickly escaped and has since killed tens of thousands of majestic California oaks. It moved to Oregon in 2001, again on nursery stock. Unbelievably, in May of 2005, infected nursery stock was shipped from California and Oregon to 23 other states. Georgia alone received 59,000 infected plants, and the public snapped up 49,000 of them before the mistake was caught (Kirby 2004). It is not hard to find a plant pathologist who thinks our oaks will go the way of the chestnut, thanks to sudden oak death disease. We can only hope they are wrong.

Alien plants beget alien insects

As with plant diseases, the list of alien insects that have been imported into North America through the nursery trade is impressive (if bad news impresses you) and includes such serious pests as Japanese beetle, cottony cushion scale, viburnum leaf beetle, citrus long-horned beetle, hemlock woolly adelgid, and balsam woolly adelgid. When it comes to homeowner annoyance, Japanese beetles win hands down. *Popillia japonica* first appeared in this country in 1916, when Henry A. Dreer, a nurseryman from Riverton, New Jersey, imported an order of Asian iris (Hansens & Weiss 1954). Henry didn't know that Japanese beetle larvae were nestled among the roots of his iris plants until they emerged as hungry adults. By then it was too late.

With no insect enemies in this country, the females laid eggs in Henry's lawn, and very little mortality curtailed the resulting larvae. The beetles spread slowly but surely westward and now threaten to cross the Mississippi. What makes Henry's oversight so destructive is that the Japanese beetle is one of the most polyphagous beetles on earth. Adults happily eat the foliage of over 400 plant species, leaving only the leaf veins. Because they favor plants from their homeland, they are particularly hard on our favorite Asian ornamentals. Japanese beetles are excellent multitaskers who enjoy eating

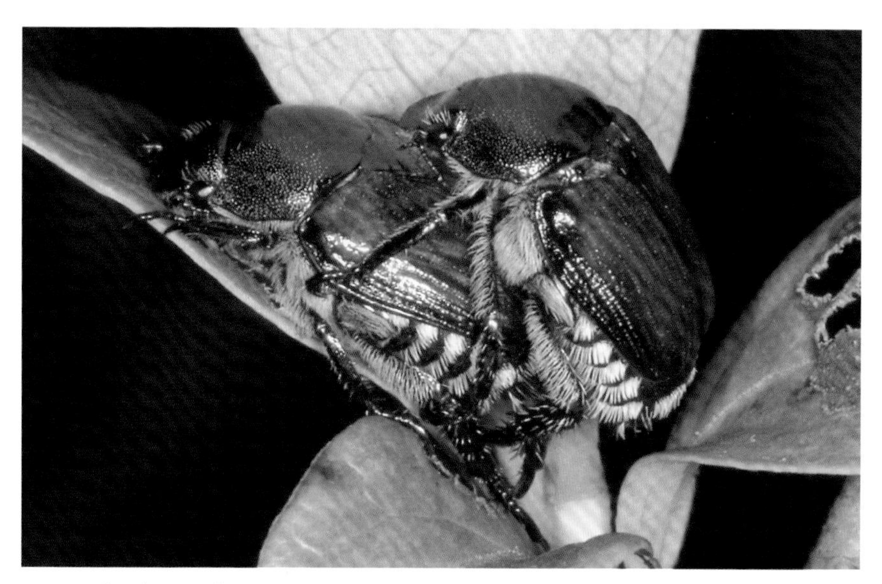

Japanese beetles (*Popillia japonica*) enjoy snacking on the foliage of over 400 plant species, even while mating.

and mating at the same time. While dining, receptive females release a very powerful pheromone that attracts all males in the area. This often results in a roiling pile of beetles, each of which takes a bite or two of the leaf they are on while trying to copulate.

Exacerbating the problem is our love of huge lawns. Japanese beetle larvae develop primarily on grass roots. With up to 40 million acres of North America covered in lawns (Milesi et al. 2005), there is no shortage of food for the beetles' young. Every year billions of adults emerge from our lawns to skeletonize our plants. One obvious way to fight this scourge is to limit the amount of food we provide for these little buggers: reduce the size of your lawn!

Lest you think that Japanese beetles are another exceptional case that would have been detected in time had they been imported on nursery stock today, think again. Three additional turf pests, all closely related to the Japanese beetle and all introduced accidentally on nursery stock after the Japanese beetle, have established themselves in the East and are beginning their westward expansion. *Exomala orientalis*, the oriental beetle (Cowles 2003), and *Maladera castanea*, the Asiatic garden beetle (Nielsen 1989), both look a lot like the Japanese beetle, although neither is as colorful. *Rhizotrogus majalis*, the European chafer, not to be confused with the native rose chafer (Nielson 1992), more closely resembles a small June beetle. As with the Japanese beetle, these three species entered the country unnoticed, despite our best intentions to screen them out.

Less annoying but far more serious in its effects on forest ecosystems is *Adelges tsugae*, the hemlock woolly adelgid (McClure, Salom & Shields 2001). Hailing from Asia, this is yet another pest that escaped detection at our ports. But who can blame the inspection agents? The hemlock woolly adelgid is a tiny relative of aphids whose eggs are far too small to be seen with the naked eye. It is absurd to think that we can catch every egg that has been laid on imported nursery stock. Hemlock woolly adelgids were first introduced to the Pacific Northwest in 1927, where they are a minor problem on western hemlocks. The real trouble started when they were shipped (on nursery stock) to Virginia in the 1950s. These insects are lethal to eastern hemlock and can kill mature trees in one to four years. They have now spread throughout the range of eastern hemlock and are threatening the

Japanese beetles produced by my neighbors' lawns completely skeletonize our Virginia creeper every year.

majestic hemlock forests of the Appalachians. As if to mock our attempts to keep these foreign creatures from entering our country on nursery stock, a second species, the balsam woolly adelgid (*Adelges piceae*), has all but eliminated the Fraser fir (*Abies fraseri*) from the high altitudes of the Great Smoky Mountains National Park (Amman & Speers 1965). This insect was brought to New England from Europe in 1900 and then was moved on nursery firs to North Carolina in the 1930s. Because Fraser firs are endemic to the Smokies (found nowhere else in the world), their loss from the area is equivalent to their extinction).

Pests imported by the ornamental industry have also had serious impacts on agricultural crops. One of the most recent examples is the soybean aphid (*Aphis glycines*), which has cost hundreds of millions of dollars in soybean yield reductions in the Upper Midwest since its introduction in 2000. The soybean aphid entered this country from China or Japan through Chicago as eggs laid on its winter host, buckthorn (Hutchinson et al. 2003). Both Chinese buckthorn (*Rhamnus utilis*) and Japanese buckthorn (*R. japonica*) were brought to this country as ornamentals. The seriousness of the soybean aphid as a threat to soybean production, as well as the invasive nature of all species of alien buckthorns, has prompted the state of Illinois to ban the further sale of these plants (Ragsdale, Voegtlin & O'Neil 2004). This legislation is certainly appropriate, but it is too late. The infestation has already spread east across the country on common buckthorn (*R. cathartica*), all the way to the Atlantic.

An even more recent invasion is threatening to put an end to the $9 billion citrus industry in Florida (Stokstad 2006). *Candidatus liberibacter* is a bacterium that causes greening disease in citrus, a deadly disease that makes fruit inedible before it kills the tree altogether. Greening disease, considered the worst citrus disease in the world, was first seen in Florida in August 2005. Unfortunately, it is unlikely to be contained because it is spread by the Asian citrus psyllid (*Diaphorina citri*), a tiny homopteran insect that arrived on infested orange jasmine, an ornamental plant that is shipped by the thousands throughout the state by discount stores. As of April 2006, greening disease had already spread to 12 counties in Florida.

I am writing about these unwanted introductions not because I enjoy condemning the nursery industry, but rather to enable gardeners to make informed decisions about which plants to use—decisions based not only on the benefits of gardening with aliens but on the costs as well. It is simply impossible to import alien nursery stock without bringing foreign diseases and insect pests to our shores, no matter how many regulatory agencies we establish to monitor such introductions. Nurserymen are not evil people who planned to eliminate the American chestnut in the East with chestnut blight (Smith 2000), or sugar pines in the West with blister rust (van Mantgem et al. 2004), or Fraser firs in the Smokies with balsam woolly adelgids (Amman & Speers 1965), or hardwood and fruit trees in the Pacific Northwest with citrus long-horned beetle (Robson 2001), or American beech with beech scale (Houston et al. 1979), or the citrus industry in Florida with Asian psyllids (Stokstad 2006), or oaks throughout the country with sudden oak death disease (Garbelotto 2004). The introduction of these organisms, particularly those that occurred in the early 1900s, were accidents that happened because we had little understanding of the consequences of introducing foreign organisms with which our native plants and animals had no evolutionary experience. An important North American ecosystem, the eastern deciduous biome, was torn asunder when the chestnut blight hitchhiked into our country with Japanese chestnuts, all because gardeners wanted to see what an alien tree would look like in their yards. Had they known what would happen as a consequence of their planting Japanese chestnuts, I am confident that no mail orders for these trees would have been placed.

I am a little more critical of our behavior today, because now we do know what can happen when we import alien plants. It has happened over and over and over again. Could it be that both the gardening public and the nursery industry consider the elimination of key species from entire ecosystems to be "collateral damage," in the parlance of the military, an undesirable but unavoidable consequence of creating beautiful gardens with desirable exotic ornamentals? I would rather think that we continue to risk importing nasty pests and deadly diseases because we, as busy gardeners, just haven't given much thought to the consequences. There is no debate when we close our borders to carriers of human diseases like SARS, mad cow disease, and avian flu virus. Why are the native plants that sustain us and our native animals less worthy of protection?

Can alien ornamentals become invasive species?

As if vanishing wildlife and the importation of diseases and pests were not enough, there is yet another problem associated with gardening with aliens. Although many gardeners insist otherwise, the plants we put in our gardens often do not stay in our gardens. When a native species sneaks off into

A

B

C

D

Alien plants that were originally brought to this country as ornamentals often become invasive over time. In many areas of the East it is hard to find a woodlot not overrun with multiflora rose (A), autumn olive (B), kudzu (C), or oriental bittersweet (D).

the natural areas surrounding our gardens, it has simply returned whence it came. But when an alien species escapes, there is a measurable chance that it will be able to grow faster and reproduce more successfully than the native vegetation in the area. Over time, more and more offspring of the escapee displace native plants in the wild. Typically it takes decades for the population of the escapees to build to the point where we start to notice that they have spread from the spots where we originally planted them (Hobbs & Humphries 1995). By that time, they have become invasive species that expand their populations at exponential rates. Unfortunately, the list of invasive species that were intentionally introduced to this country through the ornamental trade is large and growing (Reichard & White 2001), and this accounting does not even begin to number species that stowed away as seeds or small plants in the soil surrounding the sought-after ornamentals. To date, over 5000 species of alien plants have invaded the natural areas of North America.

That alien ornamentals are particularly good at outcompeting native vegetation should not come as a surprise. As we've seen, the species we import for the ornamental trade are not a random subset of plants that have evolved on other continents but instead are attractive plants selected specifically because they are particularly well defended against insects and pathogens (Tallamy 2004). How many times have you bought a plant that is advertised as being "pest free"? A plant that is "pest free" is inherently unpalatable to insects and often is not susceptible to local diseases. Because such plants do not pass the energy they capture from the sun up the food chain, they do not become functioning members of the ecosystem in which they are planted. What's more, if one escapes from your garden, it has a competitive edge over native plants that *are* eaten by a host of native insects and *are* infected by disease organisms, precisely because it is not burdened by such attacks. Indeed, while on his voyage on the *Beagle*, Darwin noticed that a number of European plants had formed near monocultures in areas of South America where for various reasons they had been introduced. He suggested that these plants might have been so successful in their new homes because they had escaped the enemies that normally attacked them. Mark Williams (1996) named Darwin's idea the "enemy release" hypothesis, and it has gained considerable support among scientists who have compared enemy loads and growth rates of aliens in their native range and in the United States. Thus, we can think of invasive species as organisms out of balance with their surrounding community.

Propagules of alien plants escape your garden with the help of animal seed dispersers or the wind. Autumn olive berries (top) are eaten by migrating birds and moved many miles from their parent plant, while Norway maple seeds (bottom) are designed to be carried many yards by the wind.

Many gardeners vigorously oppose the notion that their beloved garden plants can and do escape from cultivation. After all, the Amur honeysuckle that they tucked in the ground as a seedling and then nurtured for years is still there, right where they put it. What most gardeners do not see is the local mockingbird or migrating warbler swoop down, pluck a berry from the bush, and then fly off. The berry, of course, surrounds a seed that is genetically programmed to germinate after it has passed through the gut of a bird. In time, the mockingbird will perch somewhere nearby, perhaps in a neighbor's garden, or more likely, on the edge of the nearby woodlot, and relieve itself of the load of alien seeds in its gut. If any of those seeds germinate the following spring, the escape from your garden is complete. The migrating warbler is even more efficient at dispersing seeds. It doesn't wait around while it processes its meal of alien berries. Instead, the warbler resumes its journey south shortly after its meal of berries and may be several hundred miles away by the time the seed is defecated.

Aliens that do not produce animal-dispersed fruits or nuts rely on the wind to carry their seeds. A Norway maple produces thousands of seeds each spring, each of which bears a winglike structure that sends the seed into a helicopter spin as it descends to the ground. If there is even a slight breeze blowing when the seed detaches from the tree, it can travel a dozen yards or more before it hits the ground. Who knows how far wind-dispersed seeds can go when a strong wind blows? One spring my wife found three Japanese maple seedlings growing on our property. After searching the nearest housing development, we found what was probably the parent tree—more than half a mile away! Some aliens produce so many seeds that the sheer volume of seeds ensures that some will move considerable distances from the parent plant. Purple loosestrife is an excellent example. A single mature plant can release over 2.7 million seeds every year of its long life. Think of the seeds produced in a wetland choked by thousands of purple loosestrife plants! These billions of seeds remain viable in the soil for at least two years and can form a seed bank of immense size. Similar seed characteristics have enabled aliens like garlic mustard to invade nearly every forested acre in the eastern deciduous biome.

Are invasive species really that bad?

You may wonder why it really matters whether alien ornamentals invade roadsides, fields, woodlots, and our national parks and forests. Much has already been written about the nasty sides of invasive species, but it's worth reiterating that the large-scale replacement of native vegetation with alien plants that we are experiencing across North America is having a number of serious consequences. In addition to their impact on insect production, aliens competitively exclude, and hybridize with, native vegetation, alter the frequency of wildfires and the availability of surface or ground water, decrease the diversity of soil biota, deplete soil nutrients, degrade aquatic habitats through soil erosion, increase the competitive pressure on endangered plant species, and degrade wildlife forage (Austin 1978; Tyser 1992; Goold 1994; Randall 1996; Duncan 1997; Wilcove et al. 1998; Mack et al. 2000; Brooks et al. 2004; Butler & Cogan 2004).

Although undisturbed climax communities are far more resistant to alien invasions than disturbed areas, no place is immune. Even our most pristine national parks are under attack. The Great Smoky Mountains National Park, for example, has been invaded by over 300 species of alien plants (Miller 2003). Suffice it to say that an invasive species is far from a benign addition to our biodiversity, and we must work to avoid further invasive introductions.

But there's the rub. How do we know which ornamental aliens have the capacity to become invasive and which do not? Unfortunately, the answer to this common question depends on whom you ask. I have attended too many conferences devoted to arguing over which of our favorite ornamentals might become invasive. The fact is, it is very difficult to predict which species might become invasive should the right conditions arise. Observing how the plant behaves today can be instructive, but it can also be dangerously misleading. Many of our worst invasives did not become aggressive for many years (Hobbs & Humphries 1995). Japanese honeysuckle, for example, was planted as an ornamental for 80 years before it escaped cultivation (Silvertown 2005). No one is sure why this lag time occurs. Perhaps during the lag period, the plant is changing genetically through natural selection to better fit its new environment. Genetic changes that serve the plant well then take time to spread throughout the population. In other words, escapees from our gardens may have to adapt through evolutionary change before they can successfully outcompete the surrounding native vegetation. This takes time, and we can never be sure when or if it will happen.

If we can't be sure about a particular plant's future behavior in nature, the safe and responsible course of action is the most unpopular one: limit our use of aliens, no matter how well-behaved they seem to be. We simply can never be absolutely sure that all of the seed from a fertile plant is staying in our gardens. No doubt this message will be a tough sell to most gardeners, but the ornamental industry has already unintentionally contributed more than its share of invasives to North America (Reichard & White 2001), and the threat of creating even more problem invasives in the future is just one more good reason to reduce the percentage of aliens in our garden designs whenever possible.

There is nothing like a few good examples to help drive home an unwelcome message, as becomes clear when we look at the problems caused by just a few of the 5000 species of alien plants that now reside in the "wild" places of North America. The first has all of the traits we associate with problematic invasives. Oriental bittersweet (*Celastrus orbiculatus*) is an aggressive vine that was imported as an ornamental in the 1860s because the showy red arils that cover its seeds were (and still are) popular for flower arrangements. These same seeds, however, are also eaten by birds, and the plant quickly escaped cultivation. Today it is found from Maine to Georgia and west to Missouri. Once established, it outcompetes just about everything with which it comes in contact. It also hybridizes with its native congener, *C. scandens*. The resulting offspring are so similar to oriental bittersweet that pure *C. scandens* has all but disappeared from areas where it overlaps with *C. orbiculatus*.

Oriental bittersweet grows many yards each year and easily reaches the canopy of the tallest forest trees. As it climbs, it girdles the trunk and branches of its host tree, its leaves blocking sunlight and preventing the tree's photosynthesis. If that damage doesn't kill the tree, the sheer weight of the vine often topples even large tree trunks during wind or ice storms. I have seen tulip, sycamore, and black cherry trees three feet in diameter brought down by oriental bittersweet vines six inches thick. Perhaps its most insidious effect is on young trees during secondary succession. Like so many of our invasives, oriental bittersweet loves the full sun that characterizes old fields as they start to return to forest. The problem is that oriental bittersweet grows much faster than any of the tree species of the eastern deciduous forest, including the fast-growing tulip tree. As soon as a young tree becomes large enough, birds perch on it and deposit bittersweet seeds at its base. These soon germinate, and the vines smother the tree within a year or two. In place of the saplings of secondary succession, we see mounds of oriental bittersweet and other aliens like multiflora rose and autumn olive. In many areas of the East, aliens are blocking succession completely. Fields that were released from agriculture 25 years ago should have become young

Even places we consider protected have been invaded by alien plants. Here kudzu vines (*Pueraria montana* var. *lobata*) smother the edges of the Great Smoky Mountains National Park.

forests 30 feet tall by now; instead, they are impenetrable thickets of alien vines, and I don't see them ever escaping their alien blanket without human intervention.

Autumn olive (*Elaeagnus umbellata*) and its more western counterpart, Russian olive (*E. angustifolia*), provide two additional examples of an ornamental gone wild. Both species were introduced in the 1830s as novel Asian plants with ornamental value. Their silvery leaves, fragrant flowers, and red berries made them appealing in spring, summer, and fall. Ironically, until recently autumn and Russian olive were promoted on the basis of their environmental benefits. Highway departments in several states planted them along roadsides to stabilize soil and "feed the birds." Birds do like the berries, and every year they flock to the planted roadsides, where many are hit by speeding cars. The birds that survive spread the seeds far and wide.

Because it is one of the few nonleguminous plants that fix nitrogen, autumn olive does well in poor soils and quickly colonizes fields and disturbed sites. Though it reaches a height of only 20 feet, autumn olive is a bushy multi-stemmed tree that shades out all native competitors. Monocultures of autumn and Russian olive are a common sight in the United States. Like oriental bittersweet, the foliage of autumn olive is inedible for almost all native insect herbivores. A field rich in goldenrod, Joe-Pye weed, boneset, milkweed, black-eyed Susan, and dozens of other productive perennials supplies copious amounts of insect biomass for birds to rear their young. After it has been invaded by autumn or Russian olive, that same field is nearly sterile.

Unless you live in the mid-Atlantic states, you may not have encountered the next example of an invasive introduced by the nursery industry—at least not yet. This species has entered the United States with nursery stock three times. The first two introductions, one in Portland, Oregon, in 1890 and one in Beltsville, Maryland, in 1937, were successfully eradicated. The third introduction, in the form of seedlings sprouting in Asian rhododendron pots in a nursery in Stewartstown, Pennsylvania, in 1938, was not (Moul 1948). When the nursery owner noticed mile-a-minute weed (*Polygonum perfoliatum*) in his rhododendron pots, he transplanted it and grew it to maturity because he thought it might have value as an ornamental. When he saw the monstrous tangle of prickly stems that resulted—from which the plant gets its other common name, devil's tear-thumb—he decided it wouldn't be such a popular ornamental after all. By then, however, the lone plant had produced hundreds of seeds, the birds had eaten many of those seeds, and the damage was done.

Mile-a-minute is a classic example of a plant that experienced a long lag time before its population exploded. The Stewartstown population remained local and rather small for 40 years before it started to race across the country in the 1980s. We will never really know why mile-a-minute waited this long to disperse, but I suspect that the explosion of the white-tailed deer population in southeastern Pennsylvania had much to do with it. It has recently been confirmed that deer eat mile-a-minute seeds, which pass harmlessly through the gut and are deposited in deer scat some distance away from the

Oriental bittersweet (*Celastrus orbiculatus*) was originally brought to this country because of the ornamental value of its seeds. It rapidly covers its support tree, blocking photosynthesis and eventually destroying the tree under its own weight. It also girdles young trees, effectively strangling them to death.

A streambank completely covered with *Polygonum perfoliatum*, mile-a-minute weed (A). The plant is also known as devil's tear-thumb because its leaf petioles (B) and stems (C) are amply supplied with recurved hooks that tear more than your thumb.

source. I am willing to bet that deer are the primary source of seed dispersal because I watched dispersal in action on our own 10 acres in the span of three short years. The first year, I found a dozen or so mile-a-minute seedlings and pulled them all out with little effort. The next year, about 50 seedlings gave it a try on our property, but I got those in time as well. The third year, trillions of baby mile-a-minutes had sprouted everywhere; not just under bird perches but throughout the woods and fields. My wife and I pulled them up by the

Japanese knotweed (*Fallopia japonica*) quickly excludes all other plants as it spreads.

wagonload, but by summer's end the ones we had missed formed a dense continuous mat of vines that covered all of the vegetation within 10 feet of the ground. Zillions of new berries were produced, and I am doomed to fight mile-a-minute for the rest of my days. My guess is that it all started when our large resident deer population downed my neighbor's mile-a-minute seeds the year before and then pooped them out all over our property.

Mile-a-minute's common name describes its growth rate. Like oriental bittersweet, it grows faster than any of our local vegetation, which it quickly covers, blocking photosynthesis and bending the supporting plant into submission. Mile-a-minute is a member of the "tear-thumbs," also an accurate description of this plant's behavior. Tiny recurved hooks defend the vines and leaf petioles. One only walks through a mile-a-minute patch in shorts once. The hooks not only tear but also frequently break off in your skin, leaving you with a thousand miniscule, but very irritating, splinters. A mile-a-minute infestation renders an area out of bounds and useless. Its only redeeming feature is that it is an annual that dies at the first frost. This gives the optimistic homeowner the impression that, with a little diligence, he or she might be able to control the plant the following spring. Because mile-a-minute only recently entered its eruptive phase of invasiveness, it has not yet spread much beyond Pennsylvania, New Jersey, Maryland, Delaware, and Virginia—but it will.

My final example of an ornamental that is now designated a noxious weed because of its invasive traits is Japanese knotweed (*Fallopia japonica*). Japanese knotweed is an upright perennial that grows to 10 feet tall. Though its leaves are broad and somewhat heart-shaped, Japanese knotweed shares many of the growth habits of bamboo. It forms extremely dense thickets from which it excludes all other vegetation. It is incredibly persistent and strong: underground rhizomes send up shoots through pavement and cement with apparent ease. Japanese knotweed shoots actually grew through the floorboards of a house in England, where this species is also a problem invasive (Japanese Knotweed Alliance 1999). It is extremely difficult to kill; mowing, chopping, or tilling this plant under only seems to encourage rapid growth. Stem parts as small as your fingernail can grow into new plants with ease (Bailey et al. 1995). Introduced as an ornamental in the late 1800s, Japanese knotweed now forms monocultures along roadways, railroads, and streambeds in 36 states, including Alaska.

CHAPTER EIGHT

Creating Balanced Communities

My message in this book is a simple one. By favoring native plants over aliens in the suburban landscape, gardeners can do much to sustain the biodiversity that has been one of this country's richest assets. I have argued that native plants support and produce more insects than alien plants and therefore more numbers and species of other animals. People who accept this logic from the perspective of creating functioning ecosystems in our growing suburbs may also be alarmed by the apparent disconnect between the typical goals of a gardener, to grow beautiful undamaged plants, and my suggestion to use gardens to produce lots of insects. Yikes! Am I crazy? Maybe just a little, but not because I want suburbia to do a better job supporting the natural world. The natural world is both beautiful and full of life. Why can't our gardens reflect that? How is it that a healthy woodlot teaming with diverse insects still has plenty of healthy plants with beautiful foliage?

How to help nature control your pests

The answer, of course, is that a healthy woodlot is a collection of plants and animals—producers and consumers—that are more or less in balance. Yes, there are insect herbivores eating the plants that grow there, but keeping these herbivores in check are dozens of species of insect predators, parasites, and diseases. These, in turn, are eaten daily by the birds, amphibians, and small mammals that reside, or simply hunt, in that woodlot. I've said little so far about the decomposers in healthy communities, but they also play a vital role in keeping the community in balance. Most decomposers are insects, and they can be present in fantastic numbers, ready to recycle the nutrients in dead plants and animals for later use by the living. Decomposers are also important components of the terrestrial food chain and help provide the energy required by higher trophic levels.

In a balanced community, with rare exceptions, no one member of the food chain dominates another, and if one species in an essentially sound system does start to run rampant, it is soon brought back into equilibrium by the other members of the community. That is why all of the leaves in native forests aren't eaten by insects, why we usually don't see huge defoliation events that rage unchecked through the woods (except when an alien species like the gypsy moth is on the loose), and why a forest in balance brings us a sense of aesthetic pleasure. If you carefully inspect individual leaves in a forest, you will find that a small portion of most of them have, in fact, been eaten by insects; but the overall effect is still one of beauty, not destruction.

A plant that has fed nothing has not done its job

Somehow along the way we have come to expect perfection in our gardens: the plastic quality of artificial flowers is now seen as normal and healthy. It is neither. Instead, it is a clear sign of a garden so contrived that it is no longer a living community, so unbalanced that any life form other than the desired plants is viewed as an enemy and quickly eliminated. Today's gardeners are so concerned about the health of their plants that they run for the spray can at the first sign of an insect. Ironically, a sterile garden is one teetering on the brink of destruction. It can no longer function as a dynamic community of interacting organisms, all working smoothly to perpetuate their interac-

Healthy forests support a diversity of wildlife without ugly defoliation by insects.

tions. Its checks and balances are gone. Instead, the sterile garden's continued existence depends entirely on the frantic efforts of the gardener alone.

Late one summer I noticed that some shoots on our tomato plants had been chewed off. Closer inspection revealed, hiding amid the leaves, a large tobacco hornworm, the larval form of the sphinx moth (Sphingidae), which specializes on solanaceous plants. I could have picked it off the plant, but I'm glad I didn't; this particular larva was playing an important part in the interaction between the herbivores in our garden and their natural enemies. The larva was actually keeping the garden in balance. Lining the backside of the hornworm were three rows of fuzzy white cocoons, each containing the pupa of a hymenopteran parasitoid in the family Braconidae. These parasitoids had already fed on the internal organs of the hornworm, dooming it to an early death. I also noticed a minute pteromalid wasp inserting its eggs into many of the braconid cocoons. Not only were our tomatoes already saved from further damage from this hornworm, but had I removed the hornworm from the garden at that point I would have also removed about thirty natural enemies of any other hornworms that may have been in the garden. Our little tomato-hornworm-parasitoid community was in balance. The braconid parasitoids were keeping the hornworm numbers low, so the tomatoes, though slightly damaged, still produced plenty of fruits with no insecticide input. The pteromalid parasitoids were keeping the braconids in check, with the result that the braconids would not eliminate all of the hornworms from the garden. If they *had* eliminated the hornworms, the braconids would have also disappeared for lack of hosts. Then, if a new female hornworm adult discovered the garden, its offspring would have had no natural enemies and our tomatoes would have been doomed.

Keeping natural enemies in your garden

The part of this story that I want to stress is that there were natural enemies in the garden in the first place. Where did they come from? How did there happen to be braconid parasitoids ready to attack the first generation

A tobacco hornworm parasitized by braconid wasps, which are parasitized in turn by a pteromalid wasp (top), one of many tiny parasitic hymenopterans that attack other hymenopterans. However, were there not in the garden other sphinx moth species also parasitized by braconid wasps, like the Pandora sphinx larva (bottom), there would be no braconids to keep the tobacco hornworm in check.

of hornworms that discovered our tomatoes? What were these parasitoids eating when there were no hornworms in the garden? The answer is that the braconids were being sustained somewhere near our house on other species of lepidopteran larvae, perhaps the snowberry clearwing, the Pandora sphinx, or the great ash sphinx. These are alternate hosts for the braconids that help maintain a community of natural enemies in the area all summer long. Without a diverse array of potential hosts—insect herbivores that serve as food for various species of predators and parasitoids—there would have been no natural enemies present when the hornworm's mother first found our tomatoes. Why did we have a diversity of alternate hosts on our property? Because we have a diversity of what supports them: native plants!

The self-sustaining balance we seek in garden communities is only achieved through complexity. Sarah Stein, renowned for her foresight regarding the role gardens will have to play in future conservation efforts, recognized this years ago. She minced no words when talking about the effects of suburban landscapes on the natural world: "You can't run a supermarket on just bread, and you can't run an ecosystem on just lawn. . . . Lawns and foundation plantings are a lot simpler than the wild landscapes they replace" (Stein 1995). "Simpler" in this sense is synonymous with impoverished. Unless we establish a food web with many levels and much redundancy, just as we would find in nature, the system is likely to falter. By redundancy I mean that each trophic level should be represented by several species that do pretty much the same thing. In that way, if one species is lost from the food web, the web will not collapse, because another species is there to perform its job. The only way to build redundancy into a garden community is to start with the first trophic level, the plants. Redundancy in plants creates redundancy in the community of organisms that rely on plants for their living. This is why the monocultures we see in agricultural plantings are so notoriously unstable. Planting only one type of crop typically favors only a few types of insect herbivores—not enough to support a diverse, redundant community of predators and parasites. Under these oversimplified conditions, herbivorous insect populations typically escape the control of their natural enemies and explode. This is good for the pesticide industry but little else.

A study by Paula Shrewsbury and Mike Raupp at the University of Maryland provides a great example of how increasing the diversity of suburban landscapes can keep pest populations in check without the use of pesticides. Examining populations of *Stephanitis pyrioides*, the azalea lace bug, Shrewsbury and Raupp (2006) determined that landscapes built with many plant

Suburban landscapes (top) with both structural diversity and species diversity support more natural enemies than simplified landscapes (bottom).

species that varied in size and shape (that is, landscapes with high species and structural diversity) had populations 100 times smaller than those supported by simple landscapes built from few plant species with little structural diversity. The reason? Complex habitats contain more natural enemies of the azalea lace bug than simple habitats.

Since we usually don't witness acts of predation, it is easy to underestimate the importance of healthy and diverse communities of predators in keeping ecosystems in balance. But imagine what the world would look like without them. My wife and I experienced such a world the year we moved into our new house. The land around the house was raw from the recent construction and grading and, between the large bare spots, supported little more than mustard and ragweed. The most notable wildlife we had near our home that summer was house flies—many millions of them, it seemed. They buzzed inside and out all day long, sunned themselves on our walls, and turned every meal into a tennis match with the flyswatter. The flies came from our neighbor to the south, who maintains a large horse barn that provides all that house flies need to make many more flies. Flies beget flies from the last spring frost to the first hard freeze in the fall, and that year most of them took refuge in our house.

The following year, though, we were surprised but delighted to find only the occasional fly in our house, and today, I have to search hard to find one. Where have the flies gone? Our neighbor's barn still churns them out by the millions, but now, almost all are eaten by predators whose populations have caught up with them. Many flies are picked off in flight before they reach our house by the tree swallows that nest in our bluebird boxes. Others are eaten by spider and insect predators that are now numerous enough to do the job. One predator, in particular, that I have seen pouncing on our flies is a species of salticid jumping spider. These small but powerful spiders have learned that our windows are great hunting grounds for flies. As soon as a fly lands, a spider stalks it. When it gets within five inches, the spider makes a mighty leap and almost always lands right on the fly. Because our gardens and trees have grown since that first summer, the salticids now have plenty of places to hide and lay their eggs. They can get through the winter without desiccating—the biggest cause of winter death among arthropods—because our plants and leaf litter provide lots of moist overwintering sites. What has probably made the biggest difference, though, is that our diverse plantings support lots of alternative food sources for the spiders, so when they cannot find enough flies to eat, they do not starve. Some people might cringe at this

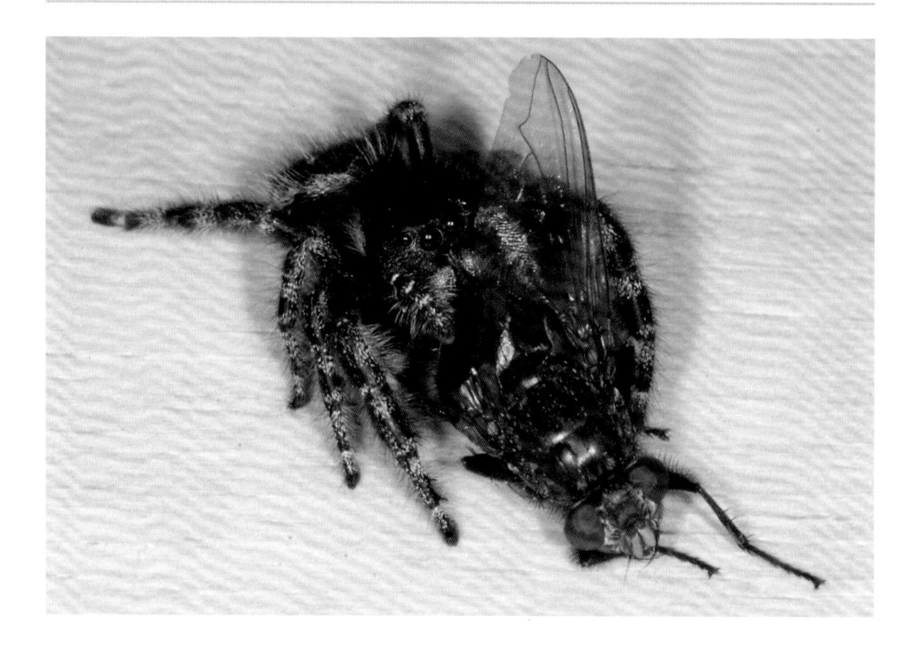

(above) The jumping spider (Salticidae) helps keep flies under control in our house.

(right) The azalea lace bug (*Stephanitis pyrioides*) is a pest from Asia that troubles azalea-lovers over much of the United States.

method of fly control, but for us, it is a good solution to an irritating problem, and we didn't have to spend a dime on traditional pest control.

Fortunately, good garden design discourages the use of only one plant species, so there is no inherent conflict between creating a beautiful garden and establishing a functioning, sustainable garden ecosystem. This brings me back to my theme of using native plants, because alien species in our gardens are often so nutritionally unavailable to the other members of the garden community we seek to establish, they might as well not be there at all. If you have 15 species of plants in your garden, and 13 of them are aliens

Predators such as the ladybird beetle (Coccinellidae) need populations of prey to survive.

that are toxic to insect herbivores, you have, in effect, a garden based on only 2 plant species. Too simple. And there's more bad news. The alien exotics that were sold to you because they were "pest free" often are anything but. It's true that they do not support insects that evolved in North America, but too many times a specialist or generalist from the alien's homeland has made the journey along with the plant and is now an established problem here. As we've seen, we can thank the ornamental plant industry for importing a great many of our worst insect pests.

In the East, the number one pest of ornamental gardens is the azalea lace bug (Raupp & Noland 1984). This bug was introduced along with evergreen azaleas from Asia and now sucks the chlorophyll from alien azalea leaves wherever the plants are massed in a sunny setting, although the bugs won't touch our native azalea species. Why don't natural enemies, the insect predators and parasites native to North America, control this pest? Because the community structure of most of our gardens is far too simple to support the numbers and diversity of natural enemies required to keep the azalea lace bug in check. Picture a classic suburban foundation planting: a row of Asian azaleas along the front of the house, bookended by two arborvitaes. Just where are the natural enemies needed to control the lace bug supposed to come from? Ladybird beetles, assassin bugs, damsel bugs, and parasitic wasps can only live in a garden if there are enough different types of prey available for them at all stages of their life cycle. Because our classic suburban foundation planting is dominated by alien plants, the only insect available to support a

A typical suburban foundation planting featuring *Rhododendron mucronulatum*, the mass-produced Asian azalea. Because this planting is too simplified and "alien" to support alternate prey, it will not attract enough natural enemies to control the inevitable outbreak of azalea lace bug.

community of predators is the azalea lace bug. When the lace bug population is small, which is the critical time for predators and parasites to prevent an outbreak, there simply is not enough prey biomass in the garden to attract and support populations of natural enemies. And so the lace bug population explodes, the homeowner runs for the insecticide, and the goal of having an undamaged garden is lost.

The same story can be told for other exotic insects that have been introduced with alien ornamental plants. Japanese beetles, as we have learned, were accidentally introduced with Asian nursery stock in the early decades of the 20th century. Japanese beetles develop on grass roots, and we have certainly given them what they want! Although there are some native insects that eat Japanese beetles, our giant lawns are typically composed of one or two alien grass species only inches high, as simplified in structure and diversity as possible. No redundancy in plant species there, and consequently no redundancy in the community of natural enemies that can survive in lawn-based habitats. The occasional predator that flies in from the woodlot down the street has no hope of controlling the millions of Japanese beetles being produced in our neighborhood lawns each year. The balance required to con-

With no healthy predator populations in the oversimplified landscapes of suburbia, Japanese beetles wreak havoc in our gardens.

trol Japanese beetles will be achieved only through well-designed landscapes founded on a diverse array of native plants. In the meantime, it's back to the insecticide to keep our "pest-free" ornamentals from being defoliated.

In an effort to create gardens free of insect problems, most gardeners have used a recipe perfect for cooking up insect outbreaks: alien plants, lack of plant diversity, insecticides. Would we not better achieve our goal of a pest-free garden if we employed nature herself to look after things? We have spent the last half-century proving beyond the shadow of a doubt that a sterile garden does not work. It is a high-input enterprise requiring more time and money than most of us would like, or are able, to devote or spend. I would love to be able to show you oodles of data proving that gardens created from native plants are more stable and self-regulating than gardens of "pest-free" aliens, but such studies have only just begun. We can wait for the results before we take action, or we can walk into the woods, look around, decide that things seem to be working pretty well there, and then do our best

Carolina silverbell (*Halesia carolina*) is a native that remains beautiful (top) even while supplying food to promethea moth larvae (bottom).

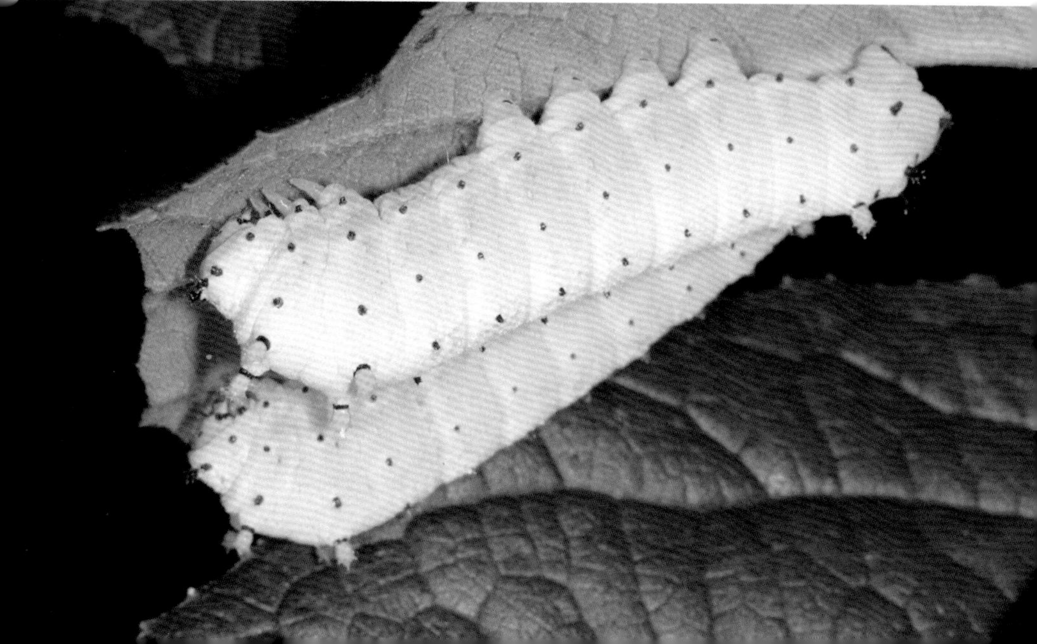

to recreate a suburban ecosystem that is complex enough to keep itself going without micromanagement on our part. To me, the choice is obvious.

One of the maxims of the new field of conservation biological control is that to control insect herbivores, you must maintain populations of insect herbivores. Although most demonstrations of this phenomenon have been in agricultural settings, the principle should work just as well in our gardens. Mike Raupp and Cliff Sadof have quantified the aesthetic tolerance limits that govern a gardener's decision to treat for insect damage (Sadof & Raupp 1996). Their study shows that as much as 10 percent of the foliage in a garden can be damaged by insects before the average gardener even notices. This is exciting news. It tells us that most gardeners do not have a zero tolerance of insects in the garden and that maybe, just maybe, the populations of insects that create 10 percent damage levels might be large enough to support communities of natural enemies so diverse and numerous that the foliage damage levels never exceed 10 percent. I believe this utopia will be easier to achieve when most plants in our gardens are native.

CHAPTER NINE

Gardening for Insect Diversity

The decision to garden with native plants in order to improve the resource base for wildlife need not be unpleasant medicine. And I hope it is not a decision that will only be popular with bird lovers. But if you delight in a golden-crowned kinglet foraging in your oak tree or an indigo bunting singing from a perch in the middle of your goldenrod patch, you may also find pleasure in the exquisite beauty of the insects that your native plants produce. Creating habitats specifically for particular insect species can be its own reward and will connect you to a fascinating part of nature that most people never meet.

Insects are worth billions!

If knowledge generates interest, and interest generates concern, it is no wonder that most people pay little heed to the plight of the insect whose native host plant has been replaced by an inedible alien. By and large, people know next to nothing about the most diverse group of organisms ever to evolve, and what they do know comes from negative encounters with a few species of biting flies, "dirty" roaches, wasps with painful stings, or crop-devouring

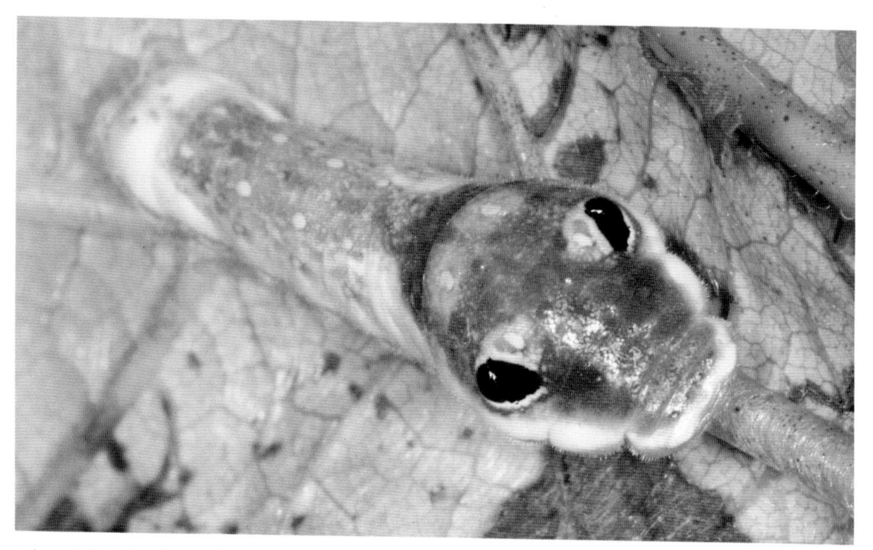

(above) Spicebush swallowtail larvae mimic the tree snakes their bird predators encounter during winters in Central America.

(left) Larvae of giant silk moths like *Hyalophora cecropia* thrive on alternate-leaf dogwood. Cecropia moths are even more likely to take up residence in your yard if you provide their favorite host plant, black cherry.

caterpillars. Our nearly universal animosity toward insects is understandable, but seriously misplaced. Of the 4 million or so insect species on earth (to put things in perspective, there are only about 9500 species of birds), a mere 1 percent interact with humans in negative ways. The other 99 percent of the insect species pollinate plants, return the nutrients tied up in dead plants and animals to the soil, keep populations of insect herbivores in check, aerate and enrich the soil, and as I keep stressing, provide food either directly or indirectly for most other animals. John Losey and Mace Vaughan (2006) have valued the ecosystem services provided by insects at $57 billion each year. As E. O. Wilson (1987) has pointed out, insects have done fine on this earth without humans and would continue to do so in our absence. If insects were to disappear, however, our own extinction would not be far behind. It may be hard to admit, but we need healthy insect populations to ensure our own survival.

Some facts about the importance of insects to life on earth may not radically change your view of these creatures, but a few minutes watch-

The viceroy butterfly (A) is more closely related to the red-spotted purple (B) than to the monarch (C).

The lace bug *Gargaphia solani* guards its eggs from predators, in this case, the larva of a green lacewing.

ing a cecropia moth crawl from her cocoon and expand her stunning wings might. Butterflies are especially attractive to people who are not otherwise entomophiles. Have you ever gazed into the "eyes" of a spicebush swallowtail caterpillar before realizing that they are not eyes at all, but part of the insect's backside? This incredible mimicry may not scare you, but the phony eyes make the caterpillar look enough like a tree snake to scare birds away. Have you realized that the bird droppings you find splattered across tulip tree leaves may actually be the larvae of the beautiful tiger swallowtail, or that the black butterflies "nectaring" at your buttonbush flowers are not the pipevine swallowtails they appear to be but rather a black morph of the tiger swallowtail that mimics the distasteful pipevine? When you see a monarch laying eggs on the milkweed plants in your yard, do you know that its offspring will fly south to the Gulf Coast, then west to Mexico, then south again to the one mountain range where they can survive the winter (Brower 1995)? Are you sure that the butterfly at your Joe-Pye weed is, in fact, a monarch; or is it a viceroy butterfly with nearly identical color patterns? Can you believe that the closest relative of the viceroy is not the monarch but the red-spotted purple, a species that looks nothing at all like the viceroy?

Lesser-known groups of insects prove every bit as interesting. The male cucumber beetle, for example, strokes his mate with his antennae in order to convince her to fertilize her eggs with his sperm and not the sperm of some other male. The tiny lace bug near its eggs on a horsenettle leaf will stand guard until the hatchlings become adults.

Gardening for butterflies

You can enjoy the fascinating world of insects if you give them the food they need to reach maturity and reproduce. The most common way gardeners attempt to connect with insects is by planting for butterflies. It is a noble idea, and exactly the sort of thing I hope to promote with this book. Sadly, the execution of this enterprise is so often directed by misinformation that we end up with fewer butterflies than we started with. When designing a butterfly garden, you need two types of plants: species that provide nectar for adults, and species that are host plants for butterfly larvae. Most people focus only on the plants that produce nectar. Even worse, they often turn to alien plants that are promoted as being good for butterflies, the most popular of which, hands down, is the butterfly bush (*Buddleja* species). Planting butterfly bush in your garden will provide attractive nectar for adult butterflies,

The alien buddleias are much touted as excellent nectar plants for butterfly gardens. Unfortunately, not a single species of butterfly in North America can reproduce on butterfly bush.

but not one species of butterfly in North America can use buddleias as larval host plants. To have butterflies, we need to make butterflies. Butterflies used to reproduce on the native plants that grew in our yards before the plants were bulldozed and replaced with lawn. To have butterflies in our future, we need to replace those lost host plants, no if's, and's, or but's. If we do not, butterfly populations will continue to decline with every new house that is built. Instead of building a butterfly garden with aliens that will make no new butterflies, use a native species that serves as a host for butterfly larvae as well as a supply of nectar for adults. This requires some knowledge, because butterflies do not lay their eggs on any old plant. They lay their eggs only on the plant species to which their larvae are adapted.

One excellent group of plants that no butterfly garden should do without are the milkweeds (*Asclepias*). When planted together, milkweed species such as butterfly weed (*A. tuberosa*), common milkweed (*A. syriaca*), and swamp milkweed (*A. incarnata*) create a continuous display of wonderful pink or orange flowers that are highly attractive to several species of butterflies from June into September. Moreover, along with the floral show, you get brand new butterflies. The monarch (*Danaus plexippus*) is the best-known

A

B

C

D

Massing species like *Asclepias syriaca*, the common milkweed (A), *A. incarnata*, swamp milkweed (B), *A. tuberosa*, butterfly weed (C), and *A. viridiflora*, green milkweed (D) will provide food for 12 species of Lepidoptera as well as a continuous bloom from June through early September.

(right) Monarch larvae develop only on milkweed species.

milkweed specialist, but at least 11 other species of Lepidoptera reproduce on milkweeds as well.

Coneflowers and black-eyed Susans (*Rudbeckia* species) also wear two hats in the butterfly garden. Along with their attractive floral display and nectar, rudbeckias support the reproduction of dozens of species of Lepidoptera, including the pearl crescent (*Phyciodes tharos*), silvery checkerspot (*Chlosyne nycteis*), and wavy-lined emerald (*Synchlora aerata*).Buttonbush (*Cephalanthus occidentalis*) is another of my favorites for the butterfly garden. Its ball-shaped flowers capture the eye, it does well in wet areas, butterflies fight to gain access to its nectar, and it serves as a host plant for 18 species of Lepidoptera in my neck of the woods. These include the ethereal promethea moth (*Callosamia promethea*), the hydrangea sphinx (*Darapsa versicolor*), and the saddleback caterpillar (*Acharia stimulea*). What better way to introduce your kids to the wonders of nature than to have them watch a female promethea moth emerge from her cocoon, inflate her wings, and without moving from her cocoon, release a powerful pheromone into the air, attracting a male of her species within minutes for coupling?

For a tall display of pink billowy flowers, use common Joe-Pye weed (*Eupatorium dubium*) or hollowstem Joe-Pye weed (*E. fistulosum*). These

The striking flower balls of *Cephalanthus occidentalis*, buttonbush, are wonderful nectar sources for butterflies in midsummer.

Rudbeckias provide a wonderful splash of color (top) as well as food for *Phyciodes tharos*, the pearl crescent (bottom).

Throughout their range, *Speyeria cybele*, the great spangled fritillary (A), and *Euptoieta claudia*, the variegated fritillary (B, larva; C, chrysalis) develop on violets (D).

If you plant buttonbush (*Cephalanthus occidentalis*) you may attract the beautiful larva of the smeared dagger moth (*Acronicta oblinita*).

(left) *Eupatorium* species (top) are a wonderful nectar source for butterflies like the tiger swallowtail and spicebush swallowtail (bottom), and they also provide food for the larvae of more than three dozen species of Lepidoptera.

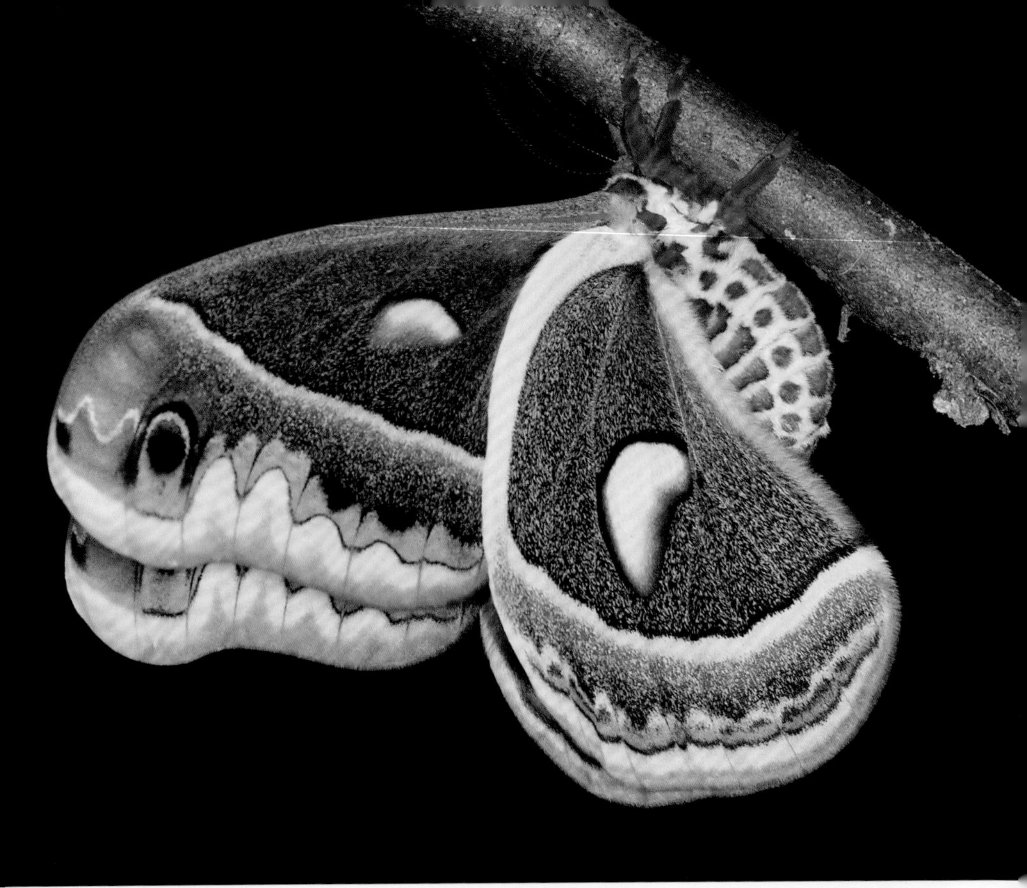

If you plant buttonbush, alternate leaf dogwood, birches, willows, or black cherry in your yard, you may attract the spectacular cecropia moth (*Hyalophora cecropia*).

species are every bit as attractive to nectaring butterflies as butterfly bush, and they host over three dozen species of Lepidoptera. Remember, every caterpillar you make in your butterfly garden becomes either a new moth or butterfly, or a source of food for a hungry bird. If you want mourning cloaks (*Nymphalis antiopa*), viceroys (*Limenitis archippus*), or io moths (*Automeris io*), plant what they eat as larvae—any of our many native willows—and you will have them. If you want to enjoy tiger swallowtails (*Papilio glaucus*), plant tulip trees, sweetbay magnolia, or black cherry trees. Black cherries will also provide food for red-spotted purples (*Limenitis arthemis astyanax*) and over 400 other moths and butterfly species.

Fritillaries eat violets (*Viola*), which can make a lovely ground cover for your butterfly garden. If you live east of Michigan, it's too late to enjoy the regal fritillary (*Speyeria idalia*). That beautiful butterfly has been all but extirpated from the East, in part by poorly timed mowing, which chops up the larvae just before they mature. But the meadow fritillary (*Boloria bellona*),

Many beautiful species such as *Isa texula*, the crowned slug caterpillar, are rarely seen—but only because most people do not take a close look at the foliage in their yards.

Diana fritillary (*Speyeria diana*), great spangled fritillary (*Speyeria cybele*), and variegated fritillary (*Euptoieta claudia*) are still doing well. Virginia creeper will enable the majestic Pandora sphinx to reproduce (by contrast, English ivy, the vine of choice in suburbia, hosts nothing). Oaks will provide food for the polyphemus moth (*Antheraea polyphemus*) and the bizarrely attractive larvae of moths in the family Limacodidae. Sweet gum (*Liquidambar styraciflua*) is your best bet for the luna moth, and it's easy to remember that the spicebush swallowtail develops on spicebush. The diminutive spring azure (*Celastrina ladon*), one of the first butterflies to become active as the days warm in the spring, may become a resident of your yard if you plant flowering dogwood (*Cornus florida*) or any of our native *Viburnum* species. They won't appear if you insist on kousa dogwood (*C. kousa*) from Asia. Azures will have nothing to do with that alien.

I could go on for pages describing the unparalleled diversity of insect life that will become an integral part of your garden when you include native host plants in your design. If you take the time to enter their world, even for a little while, these tiny marvels will not only enrich your garden, they will enrich your life as well.

The eastern tailed blue (*Everes comyntas*), a dainty lycaenid butterfly (top), develops as a larva on the spring flowers of many common woody plants. The spring azure (*Celastrina ladon*) lays an egg within the flower buds of *Viburnum dentatum* (bottom).

CHAPTER TEN

Blending In with the Neighbors

Suppose you have decided to give wildlife a helping hand by planting more natives on your property. Will your decision generate issues with the neighbors? No one wants to draw negative attention by violating community association landscaping regulations or by "letting the place go to seed." No one wants to be known as the oddball down the street who refuses to conform to neighborhood standards for yard appearance. Peer pressure from neighbors can be intense. Unfortunately, many people associate "native" plantings with front yard "prairies" planted from mail-order seeds. This, however, is not what I am advocating. I reject the notion that landscaping with natives is inherently messier or less beautiful than landscaping with aliens. In fact, because most people don't know one plant from another, I doubt if many of your neighbors would even recognize that your plant species are different from theirs.

But I do encourage one noticeable departure from traditional landscaping designs, and that is to create larger and more densely planted gardens than you may have maintained in the past. Your foundation planting could be wider, and you might add border beds that add habitat to your property while reducing the area dedicated to lawn. You may also decide to add more specimen trees or even to create a grove of trees, with an understory of small trees and shrubs appropriate to your geographic region. This more heavily planted landscape is indeed likely to contrast with the stark and barren lawns of your neighbors, but it will not be less attractive.

Lead by example

As always, knowledge is power. You may understand how important natives are to sustaining biodiversity and that adding plant biomass and diversity to your property is the key to supporting local wildlife. But in all likelihood, your neighbors do not—yet. That means it's up to you to set an example, maybe even proselytize a bit. Before you launch into your arguments, it's best to make sure you understand their perspective on landscaping. Most people have had little formal exposure to ecology. Your neighbors may have only a vague notion of what biodiversity is, and many may be skeptical of the fact that we humans cannot exist for very long on this earth without the support of healthy ecosystems. This doesn't make them bad people, but they will probably not have an intuitive appreciation for your suggestion that they consider replacing their alien butterfly bush with a native buttonbush and their English ivy with Virginia creeper. After all, aren't all plants the same? Their butterfly bush is beautiful, and look how nicely their evergreen English ivy is spreading! Don't forget to stand just beyond arm's reach when you muster the courage to tell them that if they follow your advice they too can have more insects in their yard next year!

Of course, it is unlikely that you will immediately achieve your goal of convincing your neighbors that life begins with natives (and ends with aliens). It is entirely possible that all of your neighbors will only buy into such a notion after you have shown them by example. If you construct an attractive landscape primarily from native plants, you will have begun the process of educating your neighbors' eyes to associate beauty with natives. Several years ago our good neighbor Sam (he is a good neighbor, and his name really is Sam) moseyed down our gravel driveway, stared at our rather large and certainly unconventional field of goldenrod, and asked, "Why don't you just mow that?" He was not annoyed or malicious. His question was born of genuine curiosity. Like us, Sam owns 10 acres. But, unlike us, Sam religiously maintains his property like a golf course, with tractors, commercial lawn mowers, weed whackers, and leaf blowers roaring more days than not throughout three seasons. Sam believes he is doing the right thing by keeping his property neat (and sterile) as a pin. The sight of "weeds" growing so near our house was perplexing for Sam. (Luckily, neither he nor anyone else can see our goldenrod field from their homes.)

Sam's question caught me off guard. Since beauty is in the eye of the beholder, how could I convince him that I don't mow our field because it is so lovely in the fall, with its blend of yellow goldenrod and purple New York aster? We didn't discuss how the field provides refuge from mower blades for hundreds of toads; food and nesting cover for indigo buntings, bluebirds, blue grosbeaks, and field, grasshopper, and song sparrows; a courtship arena for woodcocks in the early spring; and a place for foxes and great horned owls to catch rabbits, voles, and mice in the winter. The unmowed field also keeps the land surface from becoming so compacted by heavy machinery that it cannot absorb the hard rains that recharge our water table before the water runs off into the nearest creek. Realizing that Sam and most of the people in this country, through no fault of their own, are now so divorced from nature in their education and their everyday lives that they do not know what nature is, why we need it, or in what ways it is wonderful left me uncharacteristically tongue-tied. I mumbled something about creating habitat for wildlife and shuffled into the house. It was then and there that I decided to write this book.

After that encounter with Sam, I was convinced that I had missed a golden opportunity to explain our approach to landscaping to a tolerant and well-meaning neighbor. I should have trusted in what E. O. Wilson calls biophilia: man's innate love of nature (Kellert & Wilson 1993; Wilson 2002). Instead, I thought that Wilson didn't have Sam in mind when he conceived of his biophilia notion, and I was pretty sure that Sam thought I was nuts. My skepticism was unfounded. The following spring, I noticed two sizable patches of lawn that Sam was *not* mowing. After a few weeks, Sam stopped by and proudly announced that he was leaving those patches unmowed because "the rabbits and geese like to hide from the foxes in them!" This past summer, Sam wandered down our driveway again, this time to ask us what was eating his trees. Something had chewed his willow and river birch off at the stump and dragged the branches into his pond. Right: Sam's "golf course" had attracted a wandering beaver! Needless to say, we were ecstatic—and to our delight, so was Sam! We sat on Sam's dock with his family and watched the beaver swim back and forth until it was too dark to see. Sam wanted to know what he could feed the beaver to get it to stay (and at the same time keep it away from his specimen plantings). He was excited by this little piece of nature and he wanted to retain it in his yard. Unfortunately, acres of lawn offer little of what beavers need to make it through the seasons, so our aquatic friend moved on after two days. But the experience left me convinced that Sam, and probably most everybody, is more receptive to new ideas than I had thought. People do like living things, and after a little exposure, they may be willing to change decades-old patterns of landscaping if the payoff is more wildlife.

Native gardens need not be messy gardens

But how do we dispel this notion that native plantings are messy? Is it possible to use natives in formal settings that would please Sam? Of course! Whoever said formal, neat gardens can only be made with alien ornamentals? Some of the most beautiful gardens I have ever seen are the formal plantings of Mount Cuba Center in Greenville, Delaware. These extensive gardens almost exclusively comprise plants native to the Piedmont region of the eastern United States. Keep in mind that every one of the alien ornamentals so popular in the formal gardens of today's suburbia grows wild somewhere else in the world. Why not use natives in the same ways you would use aliens? Natives make wonderful specimen plants, mass well, and can be trained on espaliers or pruned to create formal hedges. Be as neat and formal as you want; just remember that formal does not have to mean barren.

Taking ownership of the process

One way to enlist support from your neighbors is to give them ownership of the process of saving local wildlife. Dan Janzen, a renowned ecologist at the University of Pennsylvania, has noted that humans could do much to end the extinction crisis if every person in the world became the advocate of one species. Assuming that there are about 9 million species of multicellular organisms on earth, with over 6 billion humans on the planet today, there could be more than 670 people devoted to each and every species—to monitor population levels, restore vital habitat, and lobby governments for legal protection. This is an interesting idea; but if organizing the entire world seems a bit overwhelming, why not focus on your own neighborhood? Many developments have civic associations that establish standards and rules for the homeowners in the neighborhood. If you could convince your civic asso-

Monarch butterflies are renowned for their long-distance migration to the mountain forests of central Mexico. If their forest refuges are logged, monarchs will die by the millions from exposure to winter rains and cold. Photo by Jim Plyler.

ciation to adopt one or more species that are in decline in your area, your neighborhood could collectively landscape with those species in mind.

Let's use the monarch butterfly as an example. Monarchs are in trouble for many reasons. For one thing, millions are hit by cars while en route to their Mexican wintering sites each fall. The most serious threat, though, is that the forests in which they spend the winter in the mountains of Mexico are being illegally logged (Brower et al. 2004). Thinning the forest buffer increases weather-related mortality in wintering monarch populations each year. Until the political will to protect the monarch's forests materializes, the only hope for this species is to make sure that those butterflies that do survive the winter reproduce successfully—and hugely—when they return to your backyards each summer.

A civic association could help local monarchs in several ways. First, members of the association could be encouraged to include milkweeds in private landscapes and public spaces alike. Monarchs, like most herbivores, are limited in their ability to make more monarchs primarily by food availability.

The more food provided in the form of milkweed plants, the greater the number of monarchs produced. Next, the association could coordinate the sequential cutting of some milkweed patches in June, and again in July, so that monarch larvae will have tender, young milkweed leaves to eat not only in the early weeks of summer, but in August and early September as well. The civic association might also attempt to convince the township to stop mowing roadsides that support milkweed populations during the summer. A single cutting in mid-October is enough to maintain good road visibility and will avoid the needless destruction of roadside habitat. Restraint in mowing would also save the township money, as well as reduce the production of noise, carcinogens, and climate-changing gases.

Neighborhood children could be mobilized to keep records of monarch populations over the years. What day does the first monarch return to your neighborhood each year? Whose milkweed garden is producing the most monarch larvae? How many monarchs pass through specified checkpoints in the neighborhood during the fall migration south? The degree to which your neighborhood interacts with the monarch butterfly is limited only by your imagination. I can think of no better way to reconnect with nature than to adopt a species such as the monarch, or any number of plant and bird species with declining populations. Being part of a group that successfully restores the local population of a species in trouble will not only build camaraderie with your neighbors, but may be one of the most rewarding and fulfilling things you will ever do.

CHAPTER ELEVEN

Making It Happen

If I have convinced you that there is value in having native plants in your landscape, I am pleased to have accomplished half my goal. The other half is to motivate you to take action. The more natives you incorporate into your garden, the happier the little creatures in your neighborhood will be. But philosophical issues aside, how does one change a landscape dominated by aliens to one dominated by natives? How can you achieve the beauty you have found in ornamentals from all over the world by using only local plants? Do you have to abandon what has always been your primary motivation for gardening—the creation of an artistic statement on your property—in order to recreate a landscape that supports wildlife? These are all valid—and tough—questions that will be much easier to answer 25 years from now. We are on the cusp of a paradigm shift in landscape design, and the first wave of gardeners to venture forth will truly be the pioneers who will point the way for the rest of us. Yet, while the pioneers fine-tune the use of natives in suburbia, there are a few commonsense approaches to getting started that we all can adopt right now.

How much will it cost?

Let me first dispel some fears you may have about converting your garden to natives. You need not adopt a slash and burn policy toward the aliens that are

Grouping or "massing" plants such as these alternate-leaf dogwoods (top) and river birches (bottom) creates aesthetic appeal, good habitat, and lots of bird food in the form of insects and berries.

now in your garden. Nor do you have to tolerate a lengthy transition phase as your young natives slowly assume their mature habits. I also see no reason to spend thousands of dollars on new plants. You can increase the proportion of natives in your garden in two ways, with a minimum of disruption and without destroying what is already there. First, follow the rule of attrition. When an alien dies, consider replacing it with the native species that comes closest to displaying the attributes (habit, size, texture, fall and flower color) of the lost alien. I recognize that sometimes this can be a challenge. But I predict that, as the pendulum of public interest swings toward natives, plant breeders will spend the same time and energy on enhancing cultivars of natives that they have spent on ornamental aliens over the past century. More native species will be adopted for use in the suburban garden, and replacing aliens with natives without degrading your overall garden statement will be easier than ever to achieve. In appendix 1, I offer suggestions for native landscaping plants in your region of the United States.

Start small

A second approach to working natives into your landscape is to redesign small patches of your existing garden—or, even better, create a brand-new planting on the edge of or behind an existing garden or in an area that is currently lawn. This way you can start from scratch and not have to work around old designs. Speaking of lawns: are you willing to trade in some of your lawn for specimen trees or even a small woodlot? Most yards can support many more trees than they currently do, particularly near the borders of the property, without losing the open feel near the house that people enjoy. Evolutionary psychologists believe that we humans like huge spreading lawns without any visual obstructions because we want to be able to see what danger may be out there. In the old days, we were preyed upon by lions, leopards, and hyenas, and we had to guard our territories from unfriendly neighboring clans (Wilson 1975). The desire to spot trouble early still lingers in the human psyche. Perhaps understanding what's behind our preference for large lawns will allow us to become more comfortable with less lawn and more trees. I am willing to bet that one or two more oaks in your lawn will not appreciably increase the chances that you will be set upon by mortal enemies.

Build a three-dimensional garden

Basic design concepts using natives are exactly the same as those used when landscaping with aliens. Small plants in front, tall ones in back, and so on. Remember, though, that along with a beautiful garden, you are trying to create new habitat for our animal friends. This is best accomplished by building three-dimensional gardens with a heterogeneous structure that provides places for animals to hide and nest. Native border gardens should be as wide as possible and as densely planted as possible. It's good if you can't see the ground, because then you have succeeded in providing safe sites for things that need them. Where your planting is more open, a shade-tolerant native ground cover such as mayapple, bloodroot, or eastern foamflower is appropriate. But keep in mind that simply leaving a healthy forest floor in the shadier parts of your landscape is also an option. We have been programmed to rake up every leaf and twig as it hits the ground and to try to force grass to grow where it was never meant to grow. The result is often substantial patches of bare ground.

The perfect mulch

We lose much when we remove leaf litter because it provides so many free services for us: free mulch, free fertilizer, free weed control, and free soil amendments. Litter also provides habitat for many of the arthropod predators that help keep garden communities ecologically balanced. Above all, a deep bed of leaf litter acts like a sponge, soaking up enormous quantities of water during downpours. Without litter, rainwater typically flows off our properties and into the gutters, flooding streams, rivers, and occasionally our homes. When the rain stops, leaf litter that has been allowed to accumulate slowly releases its moisture, keeping the plants and trees in your garden well hydrated, even during dry periods. Bare ground (or lawn) does none of this.

Your leaf litter can be home to a rich assortment of native plants while it is fertilizing, mulching, and watering your land. Many wildflowers grow only in soil with lots of humus, including pink and yellow lady slippers, trout lilies, bloodroot, Solomon's seal, Jacob's ladder, Jack-in-the-pulpit, numerous species of ferns and trilliums, mayapples, wild ginger, dwarf crested iris, and

A mixed grove of ash and oak trees is far more productive from a biological sense than a large expanse of lawn.

Allowing a thick layer of leaf litter to accumulate beneath shade trees enriches the soil and retains moisture even during drought.

Shaded leaf litter is the perfect habitat for dwarf crested iris, (A), bloodroot (B), Jack-in-the-pulpit (C), wild ginger (D), trout lilies (E), and mayapples (F).

A

B

C

D

F

E

You can tame natural areas of your landscape with mowed paths.

foamflower. Each of these species can add interest and beauty to what used to be hard-packed ground with sparse tufts of grass. Equally important, they create food and shelter for wildlife in areas of your property that used to be sterile.

You may worry that leaf litter will make your garden look unkempt. A good way to give your native plantings a more formal appearance is to edge them neatly. A good edging tool, of which there are many, will help to create a clear separation of garden and lawn. If you have dedicated much of your yard to tree plantings, you can lay out a well-defined path that winds its way through the grove. Even meadows can be neatened by including mowed paths and edges. Manicured paths are very effective in convincing your neighbors that your garden is planned, and not a haphazard wildland.

Chokeberry
Cranberry Bush
Bloodroot
White Pine
Hemlock Hedge
Wintercreeper
American Holly
Water Feature
Red Twig Dogwood
Iris
Sweet Pepperbush
Shed
Mountain Laurel
Juniper
Asters
Witch Hazel
Lawn
Elderberry
Phlox
Black Haw
Sweetspire
Asters
Hackberry
Nannyberry
Hydrangea
Leucothoe
Shadbush
Sourwood
Spicebush
Grey Dogwoods
Foamflower
Deck
Violets
Bearberry
Winterberry
Rhododendron
Fothergilla
White Oak
Cardinal Flower
Linden
Lawn
Carolina Silverbell
Bayberry
Cinquefoil

A design for native plantings by Ed Bruno, a landscape architect at West Chester University, West Chester, Pennsylvania. Note that the total area allocated to lawn is reduced, while an attractive and creative series of gardens planted with 38 species of ornamental natives enhances the lawn that remains.

Garden designs that fight global warming

Climatologists are now unanimous on three points: (1) carbon dioxide in the atmosphere slows the rate at which the sun's energy is lost from the atmosphere into space, and thus raises the temperature of the earth's surface; (2) there is now more carbon dioxide in the atmosphere than at any time in the past 10 million years; and (3) human consumption of fossil fuels has caused the increases in atmospheric carbon dioxide that we are currently experiencing (Kennedy & Hanson 2006). Together, these three facts suggest that it might behoove us to reduce our production of carbon dioxide. Hacking away our forests to plant huge expanses of lawn is a poor way to go about it, however. Trees are carbon sinks. That is, they use carbon from the atmosphere to build their tissues and they keep that carbon locked up and out of trouble until they die two, three, or even four hundred years later. For example, one large sugar maple tree can sequester 450 pounds of carbon dioxide each year (U.S. Department of Energy 2005). Some argue that because trees are only a temporary repository for atmospheric carbon, we should not include planting them as part of our response to global warming. It seems to me, though, that storing carbon in trees for a few hundred years might just give us enough time to come up with a more permanent solution to our carbon woes. Just think how much of the carbon emissions in the United States we could offset (and money we could save) if we reversed our course and started to replace some of our 40 million acres of lawn with trees.

When we add in the environmental costs of mowing our giant lawns, the benefits of rethinking the lawn mentality become even more obvious. If you are concerned about the human impact on our planet's climate, reducing the amount of lawn you mow each week is one of the best things you can do to reduce your family's carbon dioxide emissions. On average, mowing your lawn for one hour produces as much pollution as driving 650 miles. Moreover, we now burn 800 million gallons of gas each year in our dirty little lawnmower engines to keep our lawns at bay. In all we spend $45 billion each year on lawn care (Holmes 2006). Converting lawn to trees or garden would not only save us some money and create much needed food and habitat for our wildlife, but it would also have the twofold benefit of producing less and absorbing more carbon dioxide: a win-win endeavor.

Choose your trees wisely

If you decide to add trees and bushes to your landscape, you are halfway toward the goal of helping your local wildlife. Whether you achieve this goal or not will depend on your choice of trees. In chapter 12, I describe the 20 most valuable species of woody plants in terms of their ability to support wildlife. But you can hardly go wrong if you plant the species that originally

Many nursery dealers avoid oaks because they can be slow growing and difficult to transplant. You, however, can start your oaks from acorns. With a little pampering they grow surprisingly fast. This white oak is only five years old.

grew in your area. In our neighborhood, most lawns feature specimen trees sparsely planted here and there. About three-fourths of these are alien Bradford pears (my neighbor Sam has 29!). I suspect that this choice was made by the developer or, believe it or not, imposed on the developer by the municipality, because Bradford pears are cheap and mass-marketed. Unfortunately, they are also invasive species toxic to wildlife that try to eat them. How much better for wildlife if the developer had planted oak, beech, basswood, sweet gum, black gum, red maple, sycamore, or one of the stately hickory species. All of these can be wonderful specimen trees.

Some species, such as white oaks and hickories, are difficult to transplant once they reach a few feet tall and are therefore avoided by landscapers. The nice thing about plants, though, is that we can count on them to grow. Almost all of our native forest trees are easy to germinate from seed. All over our property my wife and I have planted dozens of native trees that we started from seed. Several post oaks have sprung up from acorns that we collected in a nearby park one weekend. Within five years, our red oaks had reached six feet, as had our alternate-leaf dogwoods, gray dogwoods, and ironwood trees. Several of our persimmon and white ash are over 12 feet tall. Our white oaks have grown more slowly but already are throwing some shade and creating their own little microhabitats. Because these trees started from seed and not from a whip or a pruned transplant, their shapes are terrific. Best of all, they were all absolutely free. In this age of instant gratification, most people do not have the mindset for starting trees from seed, but after four short years you would never know we had not spent several thousand dollars on our trees. One white oak that I did manage to move from our old house (where a squirrel had started it from seed) was two feet tall when I put it in the ground behind our new house. It is now 10 feet tall! We used to joke about having a picnic beneath it in just a few years, but the joke is on us. Soon we will spread our blanket under its branches and celebrate. And we are not the only ones who are enjoying the white oak. The tree has already produced its first polyphemus moth, a true entomological milestone.

As with any plant, you will get the fastest response from your specimen trees if you baby them. We have found that a nearby municipality collects curbside leaves each fall, composts them for a year, and then offers them free to residents who are willing to collect them. That stuff is pure gold. We call it "instant healthy forest floor" and spread it liberally beneath all of our plantings. It is the only amendment we have added to our depleted, overfarmed soil, but it holds soil moisture and seems to be all the plants need. Grass clip-

Antheraea polyphemus, the polyphemus moth (adult, top; larva, bottom) is wonderful evidence of backyard diversity.

pings may work as well, but we haven't yet tried them. When the trees are small, we also water during drought and weed out surrounding vegetation.

The value of plant diversity

When designing your native landscape, be sure to build complexity and diversity into your gardens. Creating the vertical structure found in nature's own designs by using ground cover, shrub, understory, and canopy species is a great start, but remember that animal diversity is a direct result of plant diversity. If you want to support lots of wildlife, you must supply as many different species of native plants as you can. A good place to pack in that plant diversity is in hedgerows and border plantings. You can even diversify formal front-yard plantings, particularly beneath large specimen trees. This is the perfect place for densely planted viburnums, red or black chokeberries, and native azaleas. Remember, it is the shrub layer rather than the tree canopy that birds most often use as nesting sites.

Fighting invasive species

If any part of your property borders a natural area or hedgerow, it has probably been heavily invaded by aliens such as multiflora rose, oriental bittersweet, autumn olive, Norway maple, garlic mustard, Japanese honeysuckle, Amur honeysuckle, and Japanese stiltgrass. Since this was the case on most of our property, we have learned a lot about replacing invasives with natives. The first thing we learned is that it won't happen on its own. The longer we let an area go, the more the aliens push out any existing natives. In places on our land, bittersweet had climbed to the top of host trees and was supported at the base by vines with six-inch diameters. The autumn olive trees grew four feet per year, and the multiflora rose made our house look like Sleeping Beauty's castle before the prince broke through. So armed with a small bow saw, the largest brush clippers we could buy, hand clippers, heavy gloves, and glysophate (Roundup), we set forth to do battle. After cutting or sawing through a bad guy at the base, we immediately (that means right away, not 10 minutes later), painted the stump with a Roundup solution using a small paintbrush. This is effective any time the temperature is above freezing. We chose not to spray the entire plant for fear of killing valuable native seedlings that inevitably seemed to be hiding within the alien mess. Once the plants were dead, the biggest challenge was removing the bodies. Largely

A typical hedgerow in which multiflora rose has been allowed to run amok.

for convenience, we piled them nearby as we worked; we were pleasantly surprised to see how fast these piles broke down and rotted. Most invasives are extremely fast-growers, which means they have weak wood that weathers quickly. In the meantime, the brush piles served as excellent homes and winter shelters for many local critters.

There seems to be no getting around the work involved, but the end product after removing invasives is either bare ground ready for planting or desirable natives that you have saved from untimely demise. We have been amazed at the number of native tree and bush seedlings that germinate successfully under a dense mat of aliens. We have found so many good natives under the tangle of aliens in our yard that we have sworn off mass spraying or bush-hog mowing to clear an area. Although I appreciate their efficiency, you can't use such destructive approaches without killing the natives you are trying to save. Once the mat of rose, honeysuckle, and bittersweet is cleared away, native seedlings grow quickly. If deer are a problem in your neighbor-

In our neighborhood, deer numbers are so high that we are forced to cage our young trees until they are too large for the deer to kill.

hood, you will need to place a wire cage around the plants you want to keep until they reach 5 or 6 feet. After the trees reach that height, the deer may still occasionally browse their lower branches, but that will not kill the tree; and with shrubs like *Viburnum dentatum*, deer browsing can actually give them a nice dense shape. At our house we call the day we remove a deer cage for good "Graduation Day," and it is always a celebrated event.

There are no white oaks of reproductive age within a mile of our house. This white oak seedling germinated from an acorn that was likely carried to our yard by a blue jay.

Blue jays are excellent dispersers of large seeds.

If you are faced with a large patch of bare ground after you clear away the aliens, you have two options: plant new natives yourself or let the local wildlife do it for you. Everyone knows that squirrels gather nuts (acorns, beechnuts, walnuts, and hickory nuts) in the fall and bury them for later use in the winter and spring. You may not realize, though, that blue jays do the same thing. The advantage with blue jays is that they will bring seeds to your bare patch from a good distance away, whereas squirrels tend to use only the local sources of nuts. Every year beech trees and white oaks germinate on our land, despite our having no seed-bearing beeches or white oaks on, or even near, our property. We are benefiting from forgotten seed caches of local blue jays. What's more, since jays and squirrels prefer to bury their seeds in patches of bare soil, exposing a patch of soil in the fall is all we need to do to have beech and several species of oaks and hickories germinate in the spring.

Of course, birds are also good at defecating the seeds of aliens they have eaten elsewhere, and many of those seeds will end up on your bare patch as well. Also, plants such as Japanese stiltgrass, mile-a minute weed, and garlic mustard produce so many seeds that they accumulate in the soil by the millions in what is known as a seed bank. The seed bank can remain viable for years, ready to germinate in any patch of disturbed soil. And germinate it will. We minimize the germination of aliens by heavily mulching cleared areas with our free instant forest floor (you can use weighted layers of newspaper in a pinch). The mulch also prevents soil erosion when there is a slope to the land. Nevertheless, a critical part of the restoration process is returning to your cleared area after you have planted your natives and periodically weeding out the young aliens that will surely be there. Eventually, your natives will grow to the point where they throw enough shade to slow the alien invasion. One of the great tragedies of alien introductions, however, is that the invasion will never stop altogether.

CHAPTER TWELVE

What Should I Plant?

For those who become serious about increasing animal diversity in their gardens, it is important to recognize that all native plants are not equal when it comes to supporting insect herbivores and thus other forms of wildlife. For a variety of reasons, some plant species host many dozens of specialist herbivores, while others host only a few. For example, poison ivy, ferns, and tulip trees are among the plants that few extant insect species have the ability to eat, while oaks, willows, and cherries are at the other end of the spectrum, hosting over 1400 species among them. In a study in Illinois, John Lill and Robert Marquis (2003) found that a single white oak tree can provide food and shelter for as many as 22 species of tiny leaf-tying and leaf-folding caterpillars, insects most people never notice on their walks in the woods. When all of the other lepidopterans (moths and butterflies), heteropterans (true bugs), homopterans (aphids, plant hoppers, and scales), thysanopterans (thrips), orthopterans (katydids, grasshoppers, and crickets), phasmids (walkingsticks), coleopterans (beetles), and herbivorous hymenopterans (sawflies) that develop on white oaks are considered as well, you can appreciate how important this one plant species is to the maintenance of biodiversity.

Without a doubt, the question regarding suburban restoration that I am asked most often is, "What should I plant?" The answer, of course, is to plant the species that support the most insect biodiversity. But which plants produce the most insects? You might expect every entomologist to be able to come up with a list with no trouble at all. Not so. To my knowledge, nobody has ever tried to answer this question before, and for good reasons. First of all, there are many thousands of species of insect herbivores in North America, a fair percentage of whose diets we know little or nothing about. Second, what we do know about the nutritional requirements of insect herbivores is scattered throughout several thousand scientific reports. The task of accumulating and managing all of that information would in itself be monumental. Yet the question of which plants to recommend to gardeners who want to make the biggest difference in the shortest time is so important that I am willing to compromise a bit on a comprehensive list of all host plants and let one order, the Lepidoptera, serve as a surrogate for all insect herbivores. In this case, some information is clearly better than no information.

I also am forced to slight western North America and focus on the Lepidoptera that occur on woody plants in eight states of the eastern deciduous forest biome (New York, New Jersey, Connecticut, Rhode Island, Maryland, Delaware, Pennsylvania, and Virginia). I restrict my discussion to this region because it is the only area for which we have done an exhaustive literature search for host plant relationships. Although the recommendations that follow are based on data gathered from the mid-Atlantic region, many of the plant genera discussed have broad geographic ranges and will be useful for gardeners throughout the country.

I have chosen the order of moths and butterflies as my surrogate taxon for three reasons. First, there are so many lepidopterans (11,500 species in North America) that they represent over 50 percent of all insect herbivores in this country (Arnett 2000). Coming up with some measure of host use in the Lepidoptera, therefore, should provide a good estimate of host use by all herbivores. Second, caterpillars are important components of the diets of many of the vertebrates we value most in our society, particularly birds. Finally, more information about host use in Lepidoptera has already been published than for any other large group of herbivores, making it possible to compile a list of lepidopteran host plants.

The accompanying list of woody plants ranked by their ability to support Lepidoptera species is almost entirely the result of the labors of my indefatigable research assistant, Kimberley Shropshire, who waded through host plant literature for over a year. She would be the first to point out that this list is not definitive; it is a work in progress. The number of species each plant supports is likely to change slightly as she adds to her survey over time, but I am confident that the ranking of plants based on their ability to support Lepidoptera is accurate.

WOODY PLANTS RANKED BY ABILITY TO SUPPORT LEPIDOPTERA SPECIES

Common Name	Family	Plant Genus	Species Supported
Oak	Fagaceae	*Quercus*	534
Willow	Salicaceae	*Salix*	456
Cherry, plum	Rosaceae	*Prunus*	456
Birch	Betulaceae	*Betula*	413
Poplar, cottonwood	Salicaceae	*Populus*	368
Crabapple	Rosaceae	*Malus*	311
Blueberry, cranberry	Ericaceae	*Vaccinium*	288
Maple, box elder	Aceraceae	*Acer*	285
Elm	Ulmaceae	*Ulmus*	213
Pine	Pinaceae	*Pinus*	203
Hickory	Juglandaceae	*Carya*	200
Hawthorn	Rosaceae	*Crataegus*	159
Alder	Betulaceae	*Alnus*	156
Spruce	Pinaceae	*Picea*	156
Ash	Oleaceae	*Fraxinus*	150
Basswood, linden	Tiliaceae	*Tilia*	150
Filbert, hazelnut	Betulaceae	*Corylus*	131
Walnut, butternut	Juglandaceae	*Juglans*	130
Beech	Fagaceae	*Fagus*	126
Chestnut	Fagaceae	*Castanea*	125

The following descriptions of woody plant species and the Lepidoptera that depend on them draw heavily from Dirr 1998, Sternberg & Wilson 2004, Leopold 2005, Rhoads and Block 2005, and Wagner 2005.

OAKS

Family Fagaceae, genus *Quercus*
Number of species in North America: 80

Oak trees have been a landscaping favorite for centuries, perhaps because mature specimens of so many species reflect the majestic habit of large acacias on the African plains whence we humans came. What better tree to scramble up in the event a predator should happen by? Well, that's the message our ancestors are thought to be whispering to our subconscious. Regardless of why we think they are beautiful, most of us agree that a mature oak is a desirable addition to the landscape. In fact, the oak has been named our national tree.

Most oaks fall into two taxonomic groups: the white oak group and the red oak group. How each is used in landscape designs depends on the species and the moisture level. Although all species will do well in rich, well-drained soil, pin oaks and swamp white oaks will tolerate moist soils, while scarlet oaks, chestnut oaks, and white oaks will tolerate thin, dry soils. Various species of live oaks are restricted to our southern regions, and blue oaks, Englemann oaks, and valley oaks are common west of the Sierra Nevada in California. If we disregard the diminutive scrub oak species common in pine barrens, oaks are large trees with spreading limbs when grown in full sun. A mature white oak (*Quercus alba*) can spread wider than it is tall, over 120 feet. Most oaks can be beautiful specimen trees, but massed they also make a wonderful canopy cover for woodland gardens.

The value of oaks for supporting both vertebrate and invertebrate wildlife cannot be overstated. Since the demise of the American chestnut, oaks have joined hickories, walnuts, and the American beech in supplying the bulk of nut forage so necessary for maintaining populations of vertebrate wildlife. Acorns fill the bellies of deer, raccoons, turkeys, mice, black bear, squirrels, and even wood ducks. Cavities that develop in living and dead oak giants supply vital nesting sites for dozens of species of birds, including chickadees, wrens, downy and hairy woodpeckers, flickers, owls, and bluebirds. What we have underappreciated in the past, however, is the diversity of insect herbivores that oaks add to forest ecosystems. From this perspective, oaks are the quintessential wildlife plants: no other plant genus supports more species of Lepidoptera, thus providing more types of bird food, than the mighty oak. In

Few species match the majesty of a mature white oak (*Quercus alba*).

ecological circles, oaks are noted for hosting myriad tiny moth species, mostly in the form of leaf miners and leaf tiers. But oaks are also the lifeblood for many large, showy, and sometimes positively bizarre lepidopterans. From the familiar polyphemus moth, io moth, saddleback caterpillar, and white-marked tussock moth, to the unfamiliar yellow-shouldered slug, crowned slug, and hag moth, oaks churn out caterpillars from May to October. No fewer than 20 species of dagger moths (family Noctuidae, genus *Acronicta*), 18 species of underwings (family Noctuidae, genus *Catocala*), 8 species of hairstreaks (family Lycaenidae, genera *Calycopis*, *Fixsenia*, *Satyrium*), 44 species of inchworms (family Geometridae), and 15 species of giant silk moths (family Saturniidae) prefer eastern oaks.

A careful inspection of oak leaves, particularly their undersides, often turns up caterpillars unlike any you have seen before. The crowned slug (*Isa texula*) and the smaller parasa (*Parasa chloris*) are two good examples. The crowned slug's body is ringed with defensive spines that give the appearance of beautiful jewels on a crown. The thickened body of the smaller parasa

Despite being members of the same genus, leaf shape among oaks varies widely: *Quercus stellata*, post oak (A); *Q. prinus*, chestnut oak (B); *Q. phellos*, willow oak (C); *Q. palustris*, pin oak (D).

Among the many species supported by oak are *Lithacodes fasciloa*, the yellow-shouldered slug caterpillar (A); *Parasa chloris*, the smaller parasa (B); and *Lochmaeus bilineata*, the double-lined prominent (C).

A

B

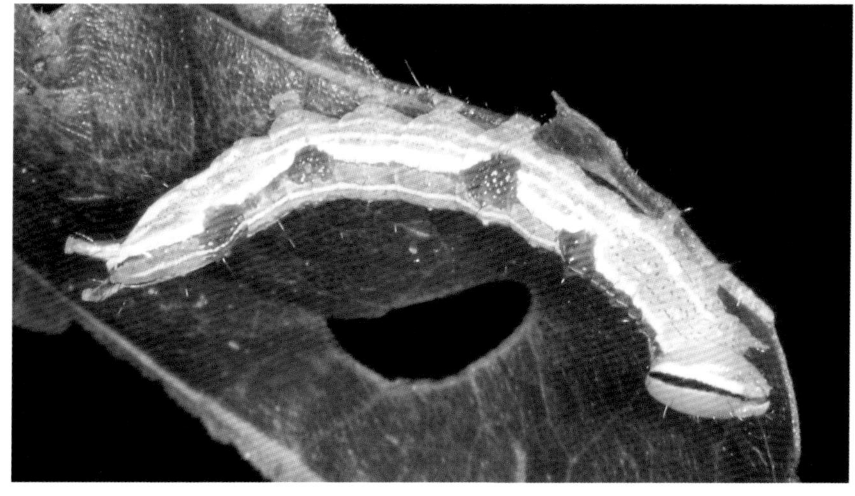

C

Many species of katydids (Orthoptera) thrive on oak foliage.

The xylem of large dead or dying oaks produces stag beetles like *Pseudolucanus capreolus* (left) and metallic wood-boring beetles like *Buprestis rufipes* (right).

looks much like a slug, which of course is a mollusk rather than an insect. Both of these species are so unlike other caterpillars that it is always somewhat startling to come across them.

Other insects use oaks for food and shelter as well. Walkingsticks and katydids mature on oak foliage, while hundreds of species of gall wasps in the family Cynipidae develop exclusively on oaks. The social bess beetles prefer oak logs for constructing their family tunnels, and oaks are a favorite of large stag beetle species throughout the United States. Unfortunately for wildlife, oaks are also excellent sources of tough wood products and so have been thinned or eliminated from far too many woodlots. Restoring large stands of oaks to suburbia would go a long way toward shoring up the future of our nation's biodiversity.

If you are planning to start your new oak trees from acorns you gather in the fall, keep in mind that acorns produced by members of the white oak group (e.g., white oak, chestnut oak, burr oak, post oak) germinate days after they fall from the tree, whereas acorns produced by red oak species germinate the following spring. Germination may not be obvious in white oak acorns unless you look for it, because only a root is produced in the fall. The rooted oak does not send up a young tree until the following spring. The fact that I can gather white oak acorns just as they start to produce their first root is one more trait that endears this tree to me. By collecting only acorns that have already germinated, I can be assured that I have viable seed that will become a young tree in the spring.

WILLOWS

Family Salicaceae, genus *Salix*
Number of species in North America: 97

Number two on the list of great Lepidoptera hosts is *Salix*, the willow. There are 97 species of willows in North America, many of which are small to medium-sized shrubs. Others are small trees, but none are notable for great size. For most of us, our experience with willows is limited to alien species such as *S. babylonica*, the weeping willow, or *S. alba*, the white willow. Mature willows are notorious for having weak wood that breaks too easily with ice, wind, or age. Our native bush species, however, are clean plants that work well as edge plantings in any damp or wet habitat. Willows have fine texture and are particularly effective when massed. If you have appropriate habitat, by all means use them liberally, for they support several of our showiest butterflies. Commas, viceroys, red-spotted purples, mourning cloaks, and several species of hairstreaks are just some of the butterflies you can attract to your yard with *Salix*. Willows also host many species of sphinx moths, all with strikingly beautiful caterpillars. The larger willow species, such as black willow (*S. nigra*) and goat willow (*S. caprea*), are also excellent hosts for several wood-boring beetles and therefore attract woodpeckers that eat the larvae of these beetles all winter long. Most nurseries do not carry native willow species, but specialty nurseries that supply materials for large restoration projects definitely do.

Black willow (*Salix nigra*) is a streamside denizen near our house that supports several species of showy butterflies.

Downy woodpeckers forage for insects in the soft wood of large willows.

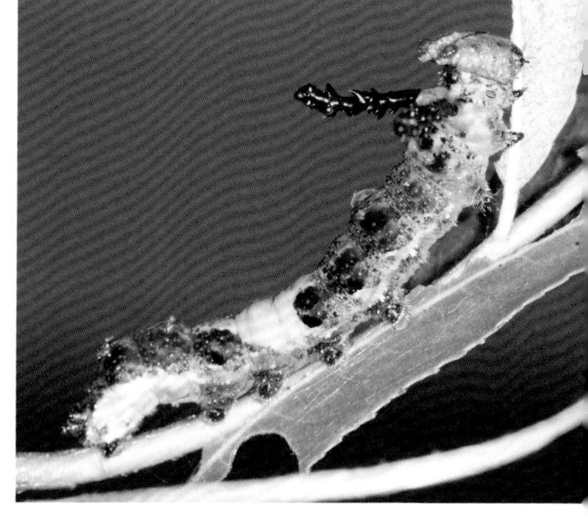

Limenitis archippus, the viceroy butterfly (left) develops as a larva (right) on willow leaves.

One of my favorite butterflies, the viceroy (*Limenitis archippus*), is most commonly found near willow hosts. Viceroys are superb mimics in all their life stages. Viceroy larvae look much like bird droppings, as do the chrysalids. The adults are the spitting image of a small monarch butterfly. For decades the viceroy's mimicry of the monarch was the textbook example of Batesian mimicry: a tasty insect copying the looks of a distasteful insect to gain protection from predators. David Ritland and Lincoln Brower (1991) have tested this hypothesis by letting redwing blackbirds taste viceroy abdomens. To Ritland and Brower's surprise, they found that, like its monarch model, the viceroy was unpalatable to birds. Viceroys are not Batesian mimics of monarchs at all, but rather Müllerian mimics. That is, the monarch and the viceroy share the same color pattern to make it easier for predators to recognize bad taste when they see it.

CHERRIES AND PLUMS

Family Rosaceae, genus *Prunus*
Number of species in North America: 31

The genus *Prunus* is an important source of food for wildlife and humans alike and has also contributed to the ornamental trade. Several species of sweet and tart cherries have been imported from Europe for human consumption, while *P. subhirtella* from China has given rise to many of the flowering cherries so popular in the ornamental trade. Plums (*P. alleghaniensis*, *P. americana*, *P. maritima*, and *P. angustifolia*) are also derived from *Prunus* stock, but are entirely of North American origins.

Native wild cherries, including black cherry (*Prunus serotina*), chokecherry (*P. virginiana*), and pin cherry (*P. pensylvanica*), are excellent sources of food for both vertebrate and invertebrate wildlife. The genus *Prunus* ranks third in the number of Lepidoptera it supports on its foliage. In the East, these include 10 species of giant silk moths, such as the cecropia moth, polyphemus moth, imperial moth, and io moth; 5 species of butterflies, such as the tiger swallowtail and red-spotted purple; 63 species of inchworms (Geometridae); and 18 species of dagger moths in the genus *Acronicta*. Cherry trees also are copious berry producers whose fruits help sustain birds for weeks in late summer.

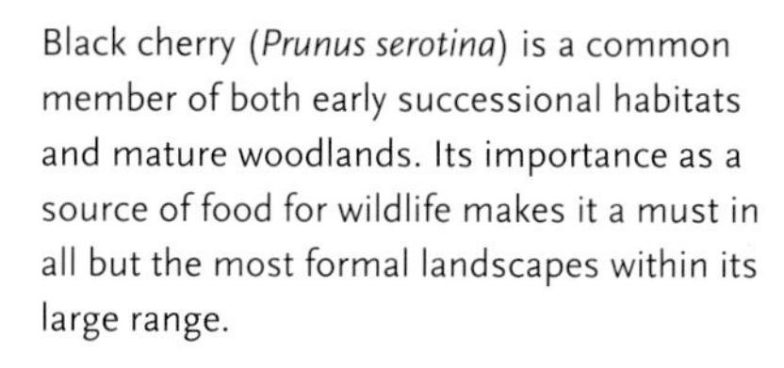

Black cherry (*Prunus serotina*) is a common member of both early successional habitats and mature woodlands. Its importance as a source of food for wildlife makes it a must in all but the most formal landscapes within its large range.

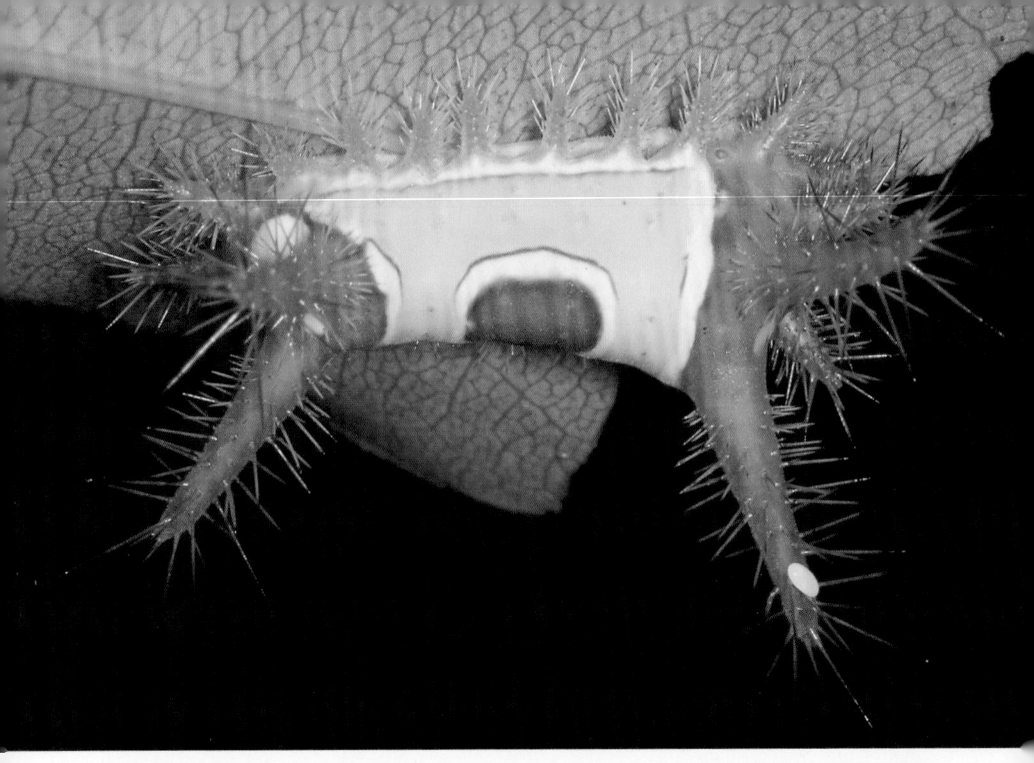

The saddleback caterpillar (*Acharia stimulea*) is a common representative of the fauna found on black cherry. Note the white tachinid parasitoid egg that has been laid on the caterpillar's dorsal horn.

One of our most striking silk moths, the cecropia moth (*Hyalophora cecropia*), prefers cherry throughout most of its range.

Cherries often produce eastern tent caterpillars (*Malacosoma americanum*), in the spring. Both larvae (top) and adults (bottom) are a favorite food of cuckoos.

Prunus americana, the American plum, is popular as a native ornamental (top). It and other *Prunus* species support *Sphinx drupiferarum*, the wild cherry sphinx (right, top).

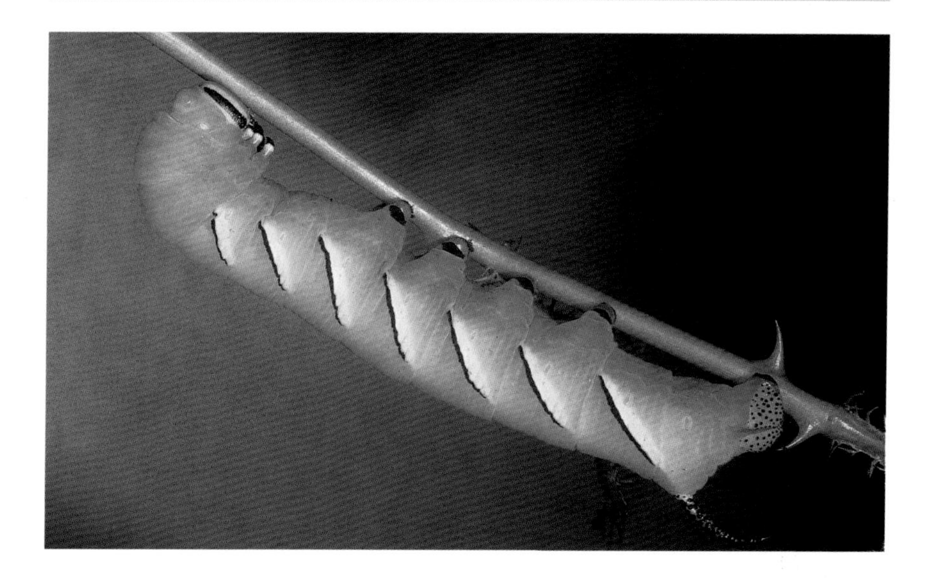

One common denizen of cherry leaves that may catch your eye (and your hand if you're not careful) is the saddleback caterpillar (*Acharia stimulea*). It is a striking member of the family Limacodidae, the slug caterpillars. With its head tucked under its stout body, it is difficult to tell which end is which. The confusion is heightened by two symmetrical horns projecting from both the front and the rear end of the caterpillar. Take note of these, for they bear stiff spines, each of which is attached to a small but potent poison gland. If you touch a spine, its tip breaks off and the surprisingly painful toxin always seems to make its way into the wound. One only knowingly touches a saddleback caterpillar once.

As with other groups, each species of *Prunus* meets a particular landscaping need. Chokecherry and pin cherry are small early successional trees that become part of the understory when larger trees grow up around them. They do well in an edge transition from lawn to woods. Black cherry also readily colonizes open spaces, but becomes a full-sized forest tree over time. Many people shy away from *Prunus* in a suburban setting because it is a favorite of eastern tent caterpillars in the early spring. Tent caterpillars cycle in abundance, and some years they seem to be everywhere. But here is where the goal of your garden becomes important. Eastern tent caterpillar is one of the main foodstuffs of both the yellow- and black-billed cuckoos, and tent caterpillar adults are eaten by many bird species, as well as by bats. If you can learn to tolerate a tent or two in the spring, your cherries and plums will provide valuable bird food all summer long.

BIRCHES

Family Betulaceae, genus *Betula*
Number of species in North America: 16

Birches are popular with both landscapers and wildlife. Birches are most numerous in colder climates and do well in a range of soil moisture, from streamside to rocky slopes. Mature trees often reach 70–80 feet in height, but most birches in suburban landscapes are smaller and make good edge transitions to plantings with larger trees. Birches are noted for their beau-

tiful exfoliating bark. Paper birch (*Betula papyrifera*), yellow birch (*B. alleghaniensis*), and river birch (*B. nigra*) in particular, are frequently planted along paths or walkways where their multiple trunks display various shades of peeling bark. Black birch (*B. lenta*) and gray birch (*B. populifolia*) do not have exfoliating bark, but mix well with sugar and red maples, particularly in piedmont or mountainous areas of the Northeast.

Birches are excellent sources of food for wildlife. Not only do they support several hundred species of moths and butterflies, they also produce

When landscaping, most people turn to *Betula papyrifera*, paper birch (left), but *B. nigra*, river birch (right), is also a superb midsized specimen tree prized for its exfoliating bark.

seeds and flower buds that are important food sources for songbirds, small mammals, grouse, and turkeys. Species with exfoliating bark provide lots of nooks and crannies in which insects hide during winter months and thus supply woodpeckers with food when they need it most. Surprisingly, few Lepidoptera eat only birch, but those that do include the arched hooktip moth (*Drepana arcuata*), and the chocolate prominent (*Peridea ferruginea*).

POPLARS, ASPENS, AND COTTONWOODS

Family Salicaceae, genus *Populus*
Number of species in North America: 8

Poplars, aspens, and cottonwoods occur in a variety of habitats throughout North America. Poplars and aspens are dominant features of northern and mountainous early successional forests, while cottonwoods are most common along riparian corridors throughout the dry western states. "Poplars" are noted for their fast growth. A hybrid between the eastern cottonwood (*Populus deltoides*) and the black cottonwood (*P. trichocarpa*), a species from the Pacific Northwest, is commonly hawked in the ornamental trade for those who seek instant gratification. The pleasure derived from these supercharged hybrids is short-lived, however, because they die just as fast as they grow. Some *Populus* species do well in wet soils and are ideal for the restoration of streamsides or low-lying sites.

Because the secondary metabolic compounds found in poplar, aspen, and cottonwood leaves are similar to those found in the willows and some birches, these three groups of plants share dozens of species of Lepidoptera. In the East, 7 species of giant silk moths (Saturniidae), 77 species of noctuid moths (Noctuidae), 7 species of sphinx moths (Sphingidae), and 10 species of butterflies (Nymphalidae and Papilionidae), among many others, all use *Populus* for larval development.

The eastern cottonwood (*Populus deltoides*) is our largest poplar (top left and right) and supports one of our most stunning caterpillars (bottom), the hourglass furcula (*Furcula scolopendrina*).

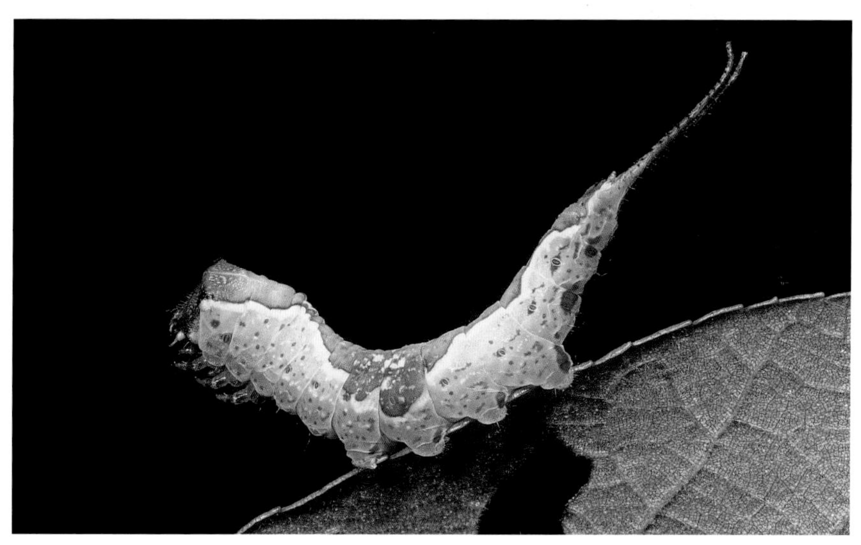

CRABAPPLES

Family Rosaceae, genus *Malus*
Number of species in North America: 4

Only four species of crabapples are native to the United States: *Malus coronaria* and *M. ioensis* are most common in the Midwest, *M. fusca* occurs in the Pacific Northwest, and *M. angustifolia* is confined to the eastern coastal plain. Flowering crabapples from Asia are so popular in the ornamental trade, however, that the *Malus* trees we most commonly encounter in our woodlots are aliens or hybrids with alien species rather than pure North American species. This is one case where the aliens are apparently so similar in leaf chemistry to our native species that there is little evidence that native insects can tell the difference, particularly when these plants hybridize. Consequently, 311 species of Lepidoptera are recorded from crabapple in the Northeast alone, including the gray hairstreak (*Strymon melinus*), the striped hairstreak (*Satyrium liparops*), 8 species of sphinx moths, and 24 moth species that eat only *Malus*. The fruits are favorite foods for many birds, deer, and other wildlife. Although not as showy as the ornamental cultivars, our natives produce a nice display of flowers in the spring and can easily stand alone as small specimen trees.

Crabapples are an excellent food source for insects and vertebrates alike.

BLUEBERRIES, CRANBERRIES, DILBERRIES, AND DEERBERRIES

Family Ericaceae, genus *Vaccinium*
Number of species in North America: 21

Blueberries, cranberries, dilberries, and deerberries are all acid-loving plants that produce copious numbers of delicious berries in mid to late summer. They are one of the first species to return after a fire, and they do well in full sun. As the forest returns around them, however, they remain as important components of the understory for decades, although berry production is minimal in full shade. Cranberries require a bog habitat with plenty of moisture.

Blueberries, dilberries, and deerberries are underused in ornamental plantings. Older specimens take on a gnarled shape that is attractive in its own right, and the flowers, fruit, and fall color give these plants landscape value in spring, summer, and fall. When massed they are excellent for wildlife, helping to nourish innumerable species of birds and mammals with their fruit. They also serve as hosts for hundreds of species of moths and of butterflies. Butterflies in the family Lycaenidae, in particular, do well on *Vaccinium* species. Henry's elfin (*Callophrys henrici*), the spring azure (*Celastrina ladon*), the brown elfin (*Callaphrys augustinus*), and the striped hairstreak (*Satyrium liparops*) all include *Vaccinium* on their larval host list.

In addition to the obvious delights the genus *Vaccinium* provides for us (left), its leaves and flowers are also a rich source of food for numerous moths and butterflies. Drexel's datana (*Datana drexelii*) is one species that specializes on blueberries (right).

MAPLES

Family Aceraceae, genus *Acer*
Number of species in North America: 9

Since the demise of the American chestnut, maples have expanded their role in forest ecosystems. Sugar maples (*Acer saccharum*) in the Appalachian Mountains and more northern reaches, mountain maples (*A. spicatum*) and striped maples (*A. pensylvanicum*) in the understory of mountain forests, silver maples (*A. saccharinum*) and box elders (*A. negundo*) on the floodplains and bottomlands, and red maples (*A. rubrum*) throughout the East all generate insect biomass from 285 species of Lepidoptera. They also supply seeds and nest sites for birds and rodents. Sugar maples, red maples, and—to the surprise of many—box elders, make excellent shade and specimen trees, while striped and mountain maples are good sources of understory foliage. Moreover, sugar maples are noted for their striking gold to orange fall foliage, while red maples add attractive yellows, oranges, reds, and purples to the fall landscape.

Acer rubrum, red maple (left), and *A. saccharum*, sugar maple (above), top the list of ornamental native maples prized both for the shade and fall color they provide.

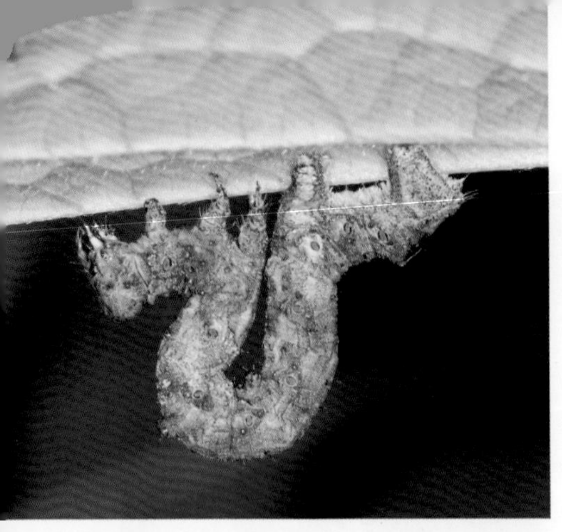

A

B

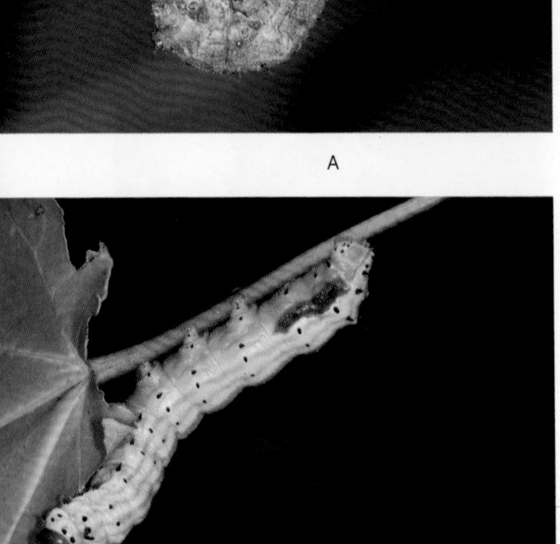

C

Maples support a diversity of inchworms, the larvae of the family Geometridae (A), as well as *Acronicta rubicoma*, the retarded dagger moth (B), and *Dryocampa rubicunda*, the green-striped mapleworm (C).

Several species of alien maples have escaped cultivation and now can be found in woodlots throughout the country. Norway maple (*Acer platanoides*) is the alien most commonly encountered in our "natural" areas. It throws such deep shade that it outcompetes native vegetation and is moving toward monoculture in many woodlots. Despite its invasive habits, Norway maple still ranks as the most commonly sold shade tree in the United States.

Maples support a wide variety of forest-loving Lepidoptera. The tree is a favorite of inchworms (Geometridae), supporting 68 species in eastern forests. As with other native plants, it provides food for several caterpillar species that eat nothing else. If you plant native maples in your yard, you are enabling the rosy maple moth (*Dryocampa rubicunda*), the oval-based prominent (*Peridea basitriens*), the retarded dagger moth (*Acronicta rubicoma*), the orange-humped maple worm (*Symmerista leucitys*), the maple looper (*Parallelia bistriaris*), and the Baltimore bomolocha (*Bomolocha baltimoralis*) to exist where they otherwise could not.

PINES

Family Pinaceae, genus *Pinus*
Number of species in North America: 35

Landscape designers rely heavily on conifers to provide accents in the winter landscape when so many other plants are drab and brown. In the northern, western, and coastal areas of the United States, several species of conifers are a natural part of the landscape. In many areas of the eastern deciduous forest, however, conifers are represented by only a few species of pines, narrowing legitimate choices of good native evergreens for this region. Nevertheless, these species include one of the most popular pines for use as majestic specimen trees or in privacy screens: the eastern white pine (*Pinus strobus*). When deep in a protected forest with rich soils, white pines can become huge, with tall, straight trunks highly valued for lumber. The current champion for Pennsylvania is 167 feet tall. In stark contrast is their appearance in the open landscape, where they often lose their leaders to ice storms, creating a spreading, flat-topped form. I have talked to homeowners who avoid white pines because of their vulnerability to the flat-topped look, but this, in fact, is exactly the classic look sought by landscapers in the know.

White pine (*Pinus strobus*), makes an excellent privacy screen while supplying food and cover for wildlife (left). When planted as a specimen tree, it typically takes on a classic flat-topped habit (right).

As one moves to the sandy, sterile soils of the coastal plain, the mountaintops, or the serpentine barrens of the East, several other species of pines become landscaping options. These include red pine (*Pinus resinosa*), pitch pine (*P. rigida*), yellow pine (*P. echinata*), and loblolly pine (*P. taeda*). Western states offer several other species with landscaping value, including ponderosa pine (*P. ponderosa*), lodgepole pine (*P. contorta*), Jeffery pine (*P. jeffreyi*), limber pine (*P. flexilis*), and bristlecone pine (*P. aristata*).

The seeds that mature within pine cones nourish turkeys, grouse, and quail; red squirrels, fox squirrels, and gray squirrels; dozens of species of song birds; and chipmunks, mice, and voles. You may not think pine needles, with all the resins and terpenes, would make a nutritious meal for moth and butterfly larvae, but they are favorites for 203 species in eastern forests. Pines host many specialists that can eat nothing else. The pine devil (*Citheronia sepulcralis*) has succumbed to development in the northern areas of its former range, but it can still be found in the pine-rich coastal plain from Virginia to Florida. Another relative of the pine devil is the imperial moth (*Eacles imperialis*). Although this species uses other trees as hosts when pine is not available, it is most common on white pine. I find the caterpillar of the imperial moth to be an exquisitely beautiful creature, unusual in that its background color is highly variable. I myself have seen green and brown individuals, but it also comes in black and red.

Both the northern pine sphinx (*Lapara bomycoides*) and the pine sphinx (*Lapara coniferarum*) are pine specialists as well. They differ in the larval stage from most other sphinx moths in not sporting a horn on the rear end. Several inchworm species also eat only pines, including the festive pine looper (*Nepytia* species), the large purplish gray (*Iridopsis vellivolata*), the Esther moth (*Hypagyrtis esther*), and the northern pine looper (*Caripeta piniata*).

A close inspection of almost any pine tree will reveal "caterpillars" that aren't caterpillars at all. They are sawfly larvae in the family Diprionidae and are more closely related to ants, wasps, and bees than to moths and butterflies. These phylogenetic realities mean nothing to hungry birds, however, who eat them as voraciously as they would any true caterpillar. It is fortunate for our local bluebirds that sawfly larvae are so common on pines. A cool wet spring often slows the development of insect populations to the point where finding enough food to feed the first clutch becomes a real challenge for bluebird parents. One spring I watched the bluebirds in my backyard rear their first clutch almost exclusively on sawfly larvae from my white pines.

Pines host the specialist *Citheronia sepulcralis*, the pine devil (A), and the more generalized *Eacles imperialis*, the imperial moth (B, C, D). They also produce large numbers of tasty diprionid sawfly larvae (E).

ELMS

Family Ulmaceae, genus *Ulmus*
Number of species in North America: 7

Elms were once important members of the deciduous forests over much of the eastern United States. When settlers first arrived from Europe, they were quick to recognize the American elm (*Ulmus americana*) as an excellent shade and specimen tree. Its graceful low-sweeping branches, broad spreading habit, fast growth, strong wood, and tolerance of compacted urban soils made it the most popular street tree in North America for more than a century. The introduction of the Dutch elm disease in 1930 abruptly ended our love affair with this tree. Few mature American elms have survived the deadly fungus. Almost as bad as losing the American elm has been our response to the loss: to import two species of Asian elms to serve as its replacement, an insult really, as Asian elms are inferior to the American elm in all traits except its resistance to Dutch elm disease. It is also ironic, for it is almost certain that the virulent fungus that causes the Dutch elm disease was originally introduced to Europe and then the United States from Asian elms. Fortunately, intense breeding programs have produced five American elm genotypes that are tolerant or resistant to Dutch elm disease: the Princeton elm, the American Liberty "multiclone," the Independence, the Valley Forge, and the New Harmony. With some luck, we might soon see the return of the American elm to the suburban ecosystem.

Elms support a great diversity of Lepidoptera, including several specialists that eat nothing else. Among these, the double-toothed prominent (*Nerice bidentata*) is the quintessential elm specialist. Not only does it eat only elm leaves, but natural selection has also shaped the back of its caterpillar stages to look just like the serrated edge of leaves of the American elm and another native, the slippery elm (*Ulmus rubra*). As it eats away the edge of a leaf, the caterpillar disguises its feeding damage with its own body. This is a handy adaptation indeed, because birds often hunt for caterpillars not by looking for the caterpillar itself, but by seeking the damage a caterpillar has left on a leaf (Heinrich & Collins 1983).

The introduction of the Dutch elm disease has made mature beauties of the American elm a rarity. This one has survived to see another spring on the campus of the University of Delaware.

Nerice bidentata, the double-toothed prominent, is so adapted to eating elm that its body resembles the edge of an elm leaf to help it hide from hungry birds.

HICKORIES

Family Juglandaceae, genus *Carya*
Number of species in North America: 12

Like many oaks, hickories are large trees that in full sun can spread as wide as they are tall. They thus make excellent specimen trees and their large compound leaves throw nice shade. Twelve species occur in North America, most east of the Mississippi. Their bark ranges from the nearly smooth trunk of the pignut hickory (*Carya glabra*) to the coarse exfoliating trunks of the aptly named shagbark hickory (*C. ovata*) and shellbark hickory (*C. laciniosa*). Hickories are noted for their production of tough nuts that are important sources of winter forage for several mammals, but they also have excellent fall color of brilliant yellows and golds. If I were a homeowner with the opportunity to convert a section of my property from lawn to trees, I would rely heavily on hickories, oaks, and beeches for my canopy species.

Hickories host many beautiful moths and butterflies. The most spectacular is the hickory horned devil, the larva of *Citheronia regalis*, the royal walnut moth. A full-grown larva often exceeds 12 centimeters (nearly 5 inches) in length and sports 10 large horns just behind the head. Though formidable

Pignut hickory (*Carya glabra*) is just one of several species that will spread to form an excellent shade tree (left). Shagbark hickory (*C. ovata*) is easily recognized by its coarse, shaggy bark (above).

Citheronia regalis, shown here in its larval (top) and adult (left) forms, is declining throughout its range and is already extinct in New England.

looking, the caterpillar is harmless. It is most often encountered when it leaves its host and wanders about in search of a pupation site that it excavates in an underground chamber. The moth that emerges the following summer, the royal walnut moth, is one of the largest and most beautiful Lepidoptera in North America.

The hag moth caterpillar (*Phobertron pithecium*) doesn't look like a caterpillar at all. Hickory is one of its favorite hosts.

As with many of our larger moths, light pollution has all but eliminated royal walnut moths from developed areas. Like sailors to a siren, moths are drawn to lights at night where they stay until the light is extinguished. Unfortunately, they usually meet an untimely end from hungry bats long before that happens. If they are not killed directly, they spend so much energy circling lights that they soon die of exhaustion. The preponderance of lights where there used to be forest is taking a heavy toll on these wonderful animals throughout their range. Royal walnut moths have already disappeared entirely from New England. (A plea: switch off unnecessary lights when you retire at night!)

Another notable lepidopteran species found on hickory is the hag moth (*Phobetron pithecium*). The caterpillar is unlike any other in this country. It bears numerous fleshy extensions of its body that are covered with dense short hairs. The total effect suggests the shed skin of a tarantula and may appear unpalatable to predators. For years, I had never seen a hag moth; then one summer I found three on my hickories. Hickories also support a number of caterpillars made fuzzy by long dense hairs. Most common among these are the hickory tussock moth (*Lophocampa caryae*), the American dagger moth (*Acronicta americana*), the contracted datana (*Datana contracta*), and the walnut caterpillar (*Datana integerrima*).

HAWTHORNS

Family Rosaceae, genus *Crataegus*
Number of species in North America: Many

The genus *Crataegus* is large and taxonomically confusing. Only a few experts in the world are capable of identifying most species with confidence, and even they can't agree on how many species there are in North America. Estimates range from 200 to 1000. Even if the minimum estimate turns out to be accurate, *Crataegus* is one of the largest genera of woody plants in the world. Like oaks, hawthorns hybridize regularly, making their identification even more challenging. All species are small trees, with many of the features that we like on crabapples (beautiful spring flowers, striking red fruits in the fall), although most species present stiff thorns up to two inches in length on their branches. When planted in full sun, their fruits can be numerous and provide nice red accents through the first several months of fall and winter.

Because of their copious fruit production, hawthorns provide food for birds and mammals alike. Their thorny branches are safe nesting sites for many bird species, and their leaves nourish 159 species of eastern caterpillars. Because hawthorns are close relatives of crabapples and cherries, they host many of the same species of Lepidoptera, including 5 sphinx moths, 10 dagger moths, and 6 butterflies. One species found on hawthorn illustrates the lengths to which some caterpillars will go to hide from birds. The blinded sphinx (*Paonias excaecatus*), snips the remains of each leaf as it eats it from the tree, thereby removing one of the primary cues birds use to locate caterpillars among the foliage.

Hawthorns are well defended by serious thorns nearly two inches long (left), but the fruits lend great fall color to any landscape (above). Photo of hawthorn fruits by John Frett.

The blinded sphinx (*Paonias excaecatus*) includes hawthorns among its larval hosts.

ALDERS

Family Betulaceae, genus *Alnus*
Number of species in North America: 8

Like many willow species, alders are multistem bushes that commonly form dense thickets on lakeshores and streamsides or in marshes. Their ability to thrive in wet sites, however, does not mean they do poorly in well-drained settings. On the contrary, alders make excellent border plantings almost anywhere within their range. The species most commonly encountered as one moves south from New England is smooth alder (*Alnus serrulata*). The gray alder (*A. incana*) is the dominant northern species.

Lepidoptera that eat birch generally like alder as well, for *Alnus* and *Betula* are closely related. Because of their affinity for wet areas, alders are excellent sources of caterpillars for marsh-breeding birds like the yellowthroat, prothonatory warbler, long-billed marsh wren, red-winged blackbird, and swamp sparrow. One moth that commonly uses alder for larval development is the banded tussock moth (*Halysidota tessellaris*). Its caterpillar is so completely covered with stiff hairs that it makes no effort to hide from predators, for some of which those hairs may make the larva more trouble than it's worth.

Alders, like this gray alder (*Alnus incana*), should be massed in wet areas or on the shores of lakes and ponds.

Alders not only host the larvae of over 150 species of Lepidoptera, such as *Halysidota tessellaris*, the banded tussock moth (top), but they also nourish numerous types of sawfly larvae like this unidentified species that was feeding in my garden (bottom).

SPRUCES

Family Pinaceae, genus *Picea*
Number of species in North America: 7

Spruces join firs and pines to form the great coniferous forests of the North and West. Blue spruce (*Picea pungens*) is our only native spruce used widely as an ornamental, although it is typically used outside its normal range. Norway spruce, on the other hand, is one of the standard alien conifers pushed by the ornamental industry in the Northeast.

Within their native range, spruces generate seeds for chickadees, nuthatches, crossbills, and pine siskins, while spruce grouse eat both the seeds and the spruce needles during the long winter months. In summer, over 150 species of moths and butterflies grow on spruce and, in turn, nourish the young of the many species of migrant birds within northern breeding grounds. Numbers of outbreak species such as the spruce budworm (*Choristoneura fumiferana*) periodically explode. Research has shown that when this occurs, the breeding success of forest birds skyrockets (Zanette & Tremont 2000). What a clear demonstration that most bird populations are limited by the amount of food they can find to feed their young.

ASHES

Family Oleaceae, genus *Fraxinus*
Number of species in North America: 16

The ashes (*Fraxinus* species) join fringe tree (*Chionanthus virginicus*), members of the genus *Forestira*, and devilwood (*Osmanthus americanus*) as the only members of the family Oleaceae native to North America. Ashes are large, globose shade trees with compound leaves and coarse-textured twigs. They are important constituents of deciduous forests throughout the United States, and several species adapt well to suburban environments. Their winged seeds nourish numerous bird and rodent species, and their foliage is host to many dozens of Lepidoptera species.

Typical of the Lepidoptera produced by ashes is the great ash sphinx (*Sphinx chersis*). A specialist on ash, the larvae reach nearly five inches before

White ash can be a stately specimen tree (top) or be massed to create good shaded habitat (bottom).

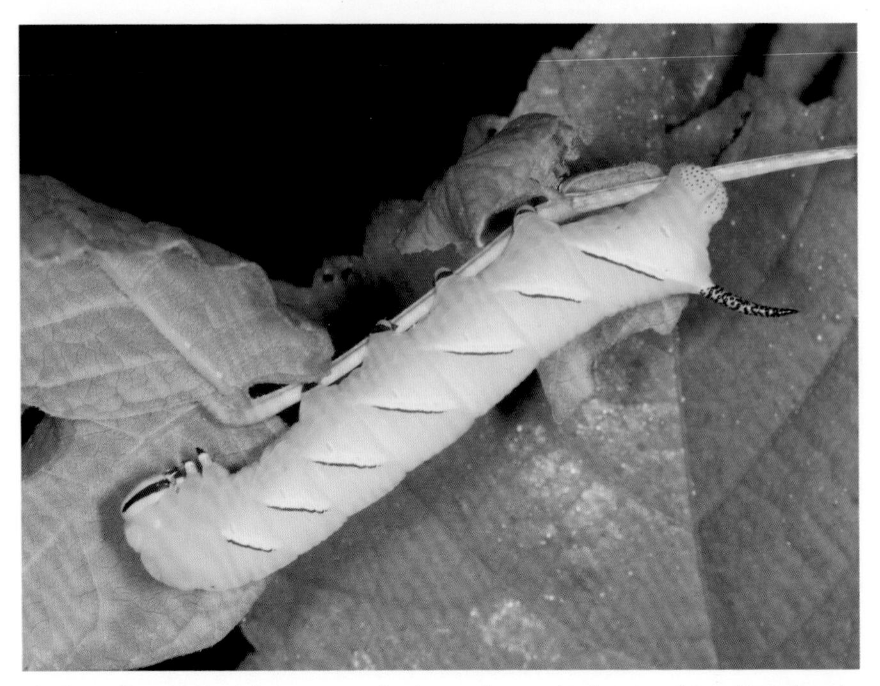

The fawn sphinx (*Sphinx kalmiae*) is one of 10 large sphinx moths that use ashes as host plants in their larval stage.

climbing off the tree and pupating in an underground chamber. A beautifully cryptic gray moth emerges two weeks later and can be seen drinking nectar with its six-inch-long "tongue" from blue lobelia, cardinal flower, trumpet honeysuckle, and other flowers with deep corollas. Ten species of sphinx moths, in addition to the great ash sphinx, use ash to complete their development. Unfortunately, ashes are in trouble. The recent accidental importation of the emerald ash borer, a metallic wood-boring beetle from Asia, threatens the future of ashes, as well as their specialized insect fauna like the great ash sphinx. The borer has already killed millions of ash trees in the Upper Midwest, and it is rapidly moving east (U.S. Forest Service 2005). As if the emerald ash borer were not enough, ash yellows disease, yet another deadly tree disease that has mysteriously appeared in North America, is killing white and green ashes throughout the Northeast (Sinclair & Griffiths 1994). The standard recommendation for diseased trees is to cut them down. What a loss.

BASSWOOD

Family Tiliaceae, genus *Tilia*
Number of species in North America: 1

Basswood, also called American linden, is a tree of bottomland soils that is confined to the eastern half of the country. It does particularly well on well-drained streambanks, but I have seen it used successfully as a spreading specimen tree, high and dry in suburbia. The taxonomy of the genus *Tilia* is in a state of flux. The current thinking is that a single species, *T. americana*, occurs in North America, of which there are two varieties or subspecies: *T. americana* var. *heterophylla*, which grows in more southern areas, and *T. americana* var. *americana*, the northern genotype.

Basswood is an excellent source of pollen and nectar for native pollinators, and its seeds are favorites of squirrels, chipmunks, and other small mammals. Its greatest contribution to wildlife, however, is the food it produces in the form of insects. *Tilia* supports over 150 species of caterpillars in North America. One of these, whose detection requires sharp eyes and careful inspection of basswood leaves, is *Schizura ipomoeae*, the checkered-

Often what appears to be a dying leaf (left) proves on closer inspection to be a tasty caterpillar hiding from birds (right). *Schizura ipomoeae*'s entire body is designed to send the false message "I am not here."

fringe prominent, once known as the morning-glory prominent because of the mistaken notion that it eats morning glories. *Schizura ipomoeae*'s body resembles a dead section of leaf and is easily overlooked by predators and caterpillar enthusiasts alike.

Another species often found on basswood leaves is the forest tent caterpillar (*Malacosoma disstria*). Unlike its cousins, the eastern and western tent caterpillars (*M. americanum* and *M. californicum*), the forest tent caterpillar does not live in a tent with its brothers and sisters. It is a solitary caterpillar that usually goes about its business unnoticed. Its body is covered with attractive soft hairs, which as a small boy I loved to feel. I can still remember wanting to make a fuzzy caterpillar coat if I could only find enough of the little guys.

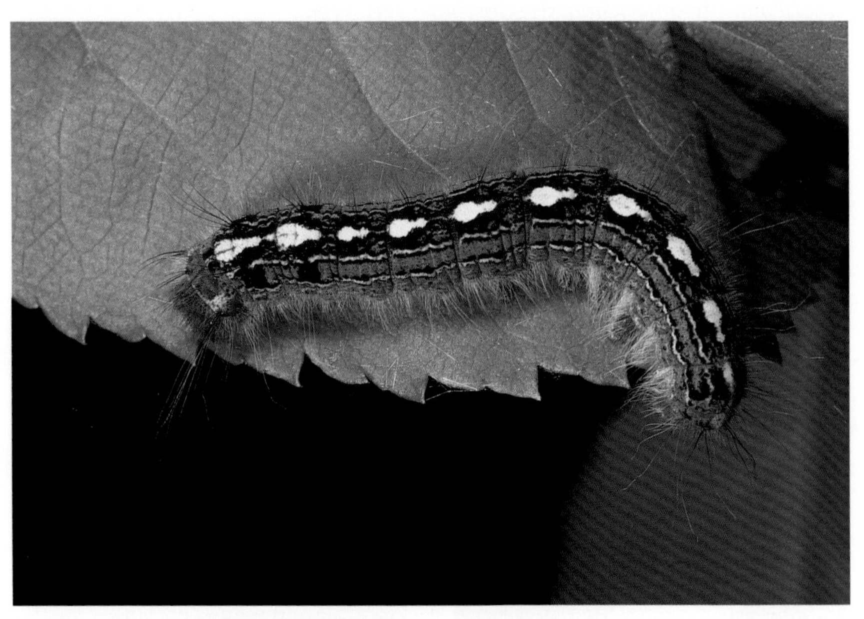

Malacosoma disstria, the forest tent caterpillar, makes no tent at all while it eats basswood leaves.

The basswood is underappreciated as a specimen tree (right) whose leaves and seeds do much to nourish local wildlife (above).

HAZELNUT

Family Betulaceae, genus *Corylus*
Number of species in North America: 2

The American hazelnut and beaked hazelnut (*Corylus americana* and *C. cornuta*), also called filberts, are common understory bushes in young, healthy woodlands. They produce delicious nuts and have been sold for that purpose since 1798. They are best massed along sunny edges where they form dense stoloniferous bushes up to 10 feet high and nut production is greatest.

American hazelnut (*Corylus americana*) is a dense understory bush (top) that produces tasty nuts (left) for both man and beast.

The larva of the white-marked tussock moth (*Orgyia leucostigma*) is one of many lepidopteran species nourished by hazelnuts.

Many Lepidoptera develop on the leaves of hazelnuts. One particularly striking example is the white-marked tussock moth (*Orgyia leucostigma*). The larvae of this species are adorned with four attractive tufts of white hairs. Few people realize that females in several species of moths do not have wings. The white-marked tussock moth is perhaps the most common of these species. Females look somewhat like a furry bag with legs. Rather than making wings they do not need, females channel all of their energy into eggs. These are deposited on the empty cocoons from which the female has recently emerged and then covered with a styrofoam-like froth for protection. Males retain their wings so they can fly to the immobile females.

WALNUTS AND BUTTERNUT

Family Juglandaceae, genus *Juglans*
Number of species in North America: 6

Juglans is a small genus of important trees with distributions in the East, California, and Texas. Black walnut (*Juglans nigra*) and butternut (*J. cinerea*) are both excellent wildlife trees because their foliage hosts well over 100 species of Lepidoptera "bird food" and because they produce large nuts that sustain squirrels and other rodents through the long winter months. Because squirrels occasionally forget exactly where they have buried *Juglans* nuts, both species spread quickly from large parent trees. Among the many species of Lepidoptera that prefer host plants in the Juglandaceae, several, including the gregarious walnut caterpillar (*Datana integerrima*), Angus's datana (*Datana angusii*), and the gray-edged bomolocha (*Bomolocha madefactalis*), are specialists that can eat nothing else.

Landscapers don't like to feature black walnut or butternut in prominent sites that will be viewed up close because of their coarse habit, and also because they are one of the first trees to lose their leaves in the fall and one of the last to grow new leaves in the spring. They are best used in informal settings mixed with other hardwoods. You may also hear dire warnings about the toxicity of walnut allelochemicals to nearby plants. Walnut tissues, particularly the husks of the nuts themselves, produce juglone, a chemical that can stunt growth or even kill other plants. The most sensitive plants, however, are alien ornamentals, because they have no evolutionary experience with juglone. Most natives that evolved within the range of walnuts are unfazed by their allelochemicals.

In recent years butternut has been marginalized in eastern forests by a serious canker disease caused by the fungus *Sirococcus clavigignenti-juglandacearum* (Ostry et al. 1997). All signs point to an Asian origin for this fungus. Relatively benign to Asian species of *Juglans*, it has spread quickly and is highly virulent in North America. We will probably never know which shipment of plants from Asia brought the fungus to the United States, but it doesn't matter for the native butternut. Its days as a contributing member of the eastern deciduous forest are numbered.

Walnuts (above top and left) are important food sources for wildlife. Many insects, like the butternut woolyworm (*Eriocampa juglandis*) and the walnut sphinx (*Amorpha juglandis*), eat nothing else (right).

AMERICAN BEECH

Family Fagaceae, genus *Fagus*
Number of species in North America: 1

Like basswood, the American beech (*Fagus grandifolia*) is the only member of its genus in North America, yet it supports over 100 species of Lepidoptera. On a per-species basis, this makes beech more productive in its support of wildlife than even the number-one-ranked oaks. Both oaks and beech belong to the plant family Fagaceae and therefore share many of the same insect herbivores. Beech also shares with the oaks the role of producing high-protein seeds that are critical components of the diet of numerous rodents, black bear, deer, raccoons, and especially, large birds like turkey and grouse.

The American beech is a denizen of mature forests. It is noted for its smooth blue-gray bark, which begs to have initials carved in it. Beech is one of the few tree species that can germinate and grow well in deep shade. In a woodlot setting it mixes well with oaks and hickories, but it also makes a magnificent specimen tree, particularly in a partially shaded setting. Like so many of our prized forest species, beech is under attack from disease. In this case, the causal agent is a native fungus but the vector is a scale insect (*Cryptococcus fagisuga*) that was brought to this country on nursery stock of European beech. Together they cause beech bark disease and are killing beech trees from New England to Philadelphia.

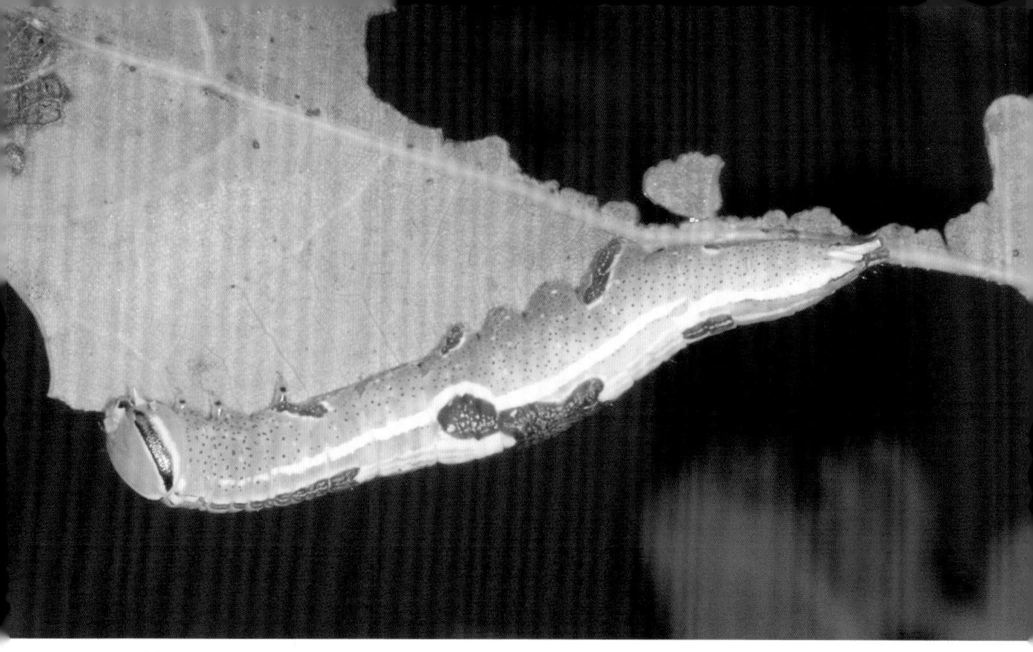

The saddled prominent (*Heterocampa guttivitta*) is a common resident of the American beech.

The smooth gray bark of *Fagus grandifolia*, the American beech (left), in combination with its distinctive foliage (right>), adds an attractive accent to woodland plantings.

CHESTNUTS

Family Fagaceae, genus *Castanea*
Number of species in North America: 2

There are three species of chestnuts in North America. The American chestnut (*Castanea dentata*) was the most common tree in the Piedmont and Appalachian regions of the East until the chestnut blight, brought to this country on resistant Japanese chestnuts, reduced our great chestnut forests to rare stump sprouts. Chinquapin (*C. pumila*) is a bush or small understory tree that was less affected by the chestnut blight and still survives in southern Pennsylvania, Virginia, and states to the south. The Ozark chinquapin (*C. ozarkensis*) is more treelike in habit than *C. pumila* and is restricted to the southern states bordering the Mississippi River. All three species produce edible nuts that sustain wildlife. Despite being too reduced to play much of an ecological role in our forests, the American chestnut still supports at least 125 species of Lepidoptera.

The American Chestnut Society is an active group with chapters in several eastern states. The primary goal of this organization is to support programs designed to breed resistance to chestnut blight in *Castanea dentata*. The hope, of course, is to replant deciduous forests with chestnuts that will not succumb to the disease. The most successful breeding program involves crossing the American chestnut with resistant Asian species and then backcrossing the offspring with the American chestnut until all Asian genes except those responsible for blight resistance are eliminated from the genotype. I am excited to report that this herculean effort is nearly complete; within a few years, resistant seed of the American chestnut should be available for planting. Fortunately, the American chestnut has great ornamental value, and I look forward to the time when it replaces Norway maple as the most popular shade tree in the country.

All that is left of the mighty chestnut forests in the East are short-lived stump sprouts.

CHAPTER THIRTEEN

What Does Bird Food Look Like?

I grew up near a small lake in northern New Jersey that was surrounded by more than a thousand acres of mature forest. My friends and I split our summertime between fun in the water and fun in the woods. Most days we were the only ones in the woods because relentless attacks from horse flies and deer flies (biting flies in the family Tabanidae) easily chased away more sensible people. Our only defense was to swat them as they attempted to drink our blood. They all looked pretty much alike to us, and we flattened them by the thousands. Years later I decided to do my master's degree research at that same lake on those same tabanid populations. I learned lots about the little guys: their names (incredibly, there were 52 different species of tabanids at that lake, two of which had never been collected in New Jersey before); what time of day they were active; and where they liked to hunt, mate, and lay eggs. I learned that only females take a blood meal and that they need the protein in that meal to develop their eggs. I learned that so many species could live in the same place at the same time from the same resource (mammal blood) because each of the 52 species had found a different way to divide the available niches at the lake. I also learned

My research on deer flies turned species like *Chrysops cincticornis* from foe to friend.

that if you let *Chrysops cincticornis* take a full blood meal from your big toe, you won't be able to get your shoe back on for an hour. Today, instead of swatting any tabanid that lands on me, I scoop it up in my hands, determine which species has come to visit, and greet my old friend (OK, sometimes I still swat them before I greet them).

I tell this story to make a simple point. Knowledge generates interest, and interest generates compassion. My master's experience taught me that if I invested some effort in understanding nature and its various components, I would no longer feel compelled to squash it as soon as it inconvenienced me. Because I discovered this with a group of insects that are very good at inconveniencing people, my hope is that providing some basic information about the benign insects that are food for birds and other wildlife will create a concern you may not have felt before for preserving these arthropods and the plants they eat.

Unless otherwise noted, the facts and figures used in the following descriptions come from Arnett 2000, Triplehorn & Johnson 2005, and Wagner 2005.

ARTHROPOD HERBIVORES

Grasshoppers and locusts

Order Orthoptera, family Acrididae
Number of species north of Mexico: 600
Life stages: egg, nymph, adult

Grasshoppers, or more accurately, short-horned grasshoppers (those whose antennae are much shorter than the body), are true locusts: the familiar insects with powerful hind legs that enable them to jump from their enemies. They frequent meadows and fields, where they develop on grasses and forbs. Grasses are well defended by silica and lignins that make their tissues tough and difficult to chew. To counter these defenses, grasshoppers have developed large mandibles (jaws, which, like the teeth of horses, are worn down over time by their diet). Grasshopper eggs are laid in the ground, where they rest during the winter and hatch the following spring.

Some grasshopper species are quite large. The lubber grasshopper of the South is a good example, often reaching over three inches in length. Lubber grasshoppers are brightly colored with red and black patterns, an advertisement that they have sequestered nasty chemicals in their tissues that make them taste terrible. But this is an exception among grasshoppers; most species taste good and are rich sources of protein and fats for birds, rodents, raccoons, and possums. In fact, indigenous peoples all over the globe include grasshoppers in their diet whenever possible. You might think our lawns would be good habitat for grasshoppers, but mowing schedules, pesticides, and the alien grasses that suburban lawns comprise prevent these important insects from completing their life cycle.

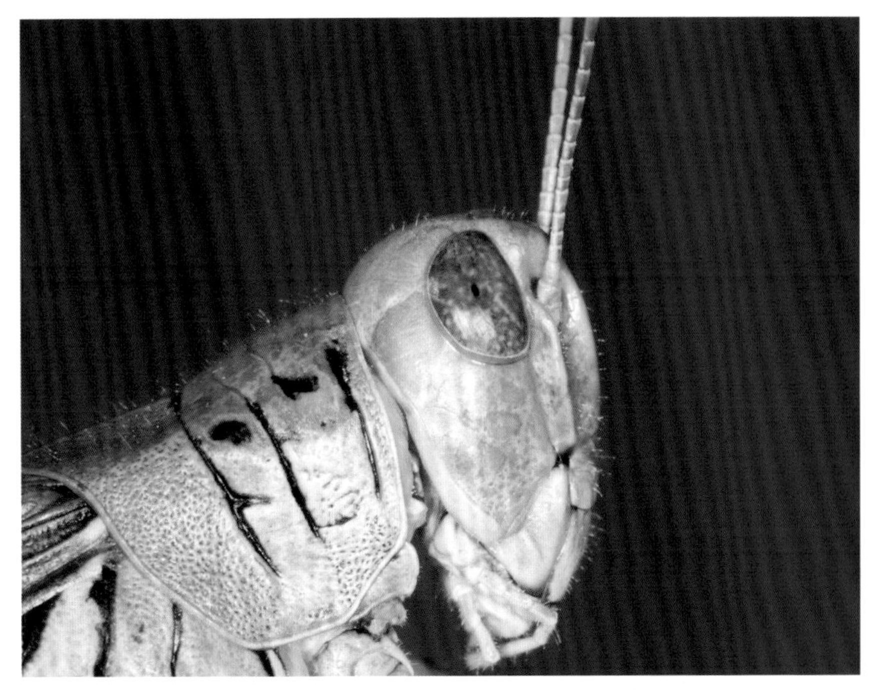

Powerful mandibles enable members of the Acrididae family to digest tough grasses and forbs.

Grasshoppers are an abundant arthropod and nourish everything from rodents to hawks.

Katydids

Order Orthoptera, family Tettigoniidae
Number of species north of Mexico: 123
Life stages: egg, nymph, adult

Katydids are also called long-horned grasshoppers because they sport antennae at least as long as their bodies and often much longer. Some of them are quite large and have wings that are the spitting image of leaves. Though some common field species are active during the day, most katydids are nocturnal and spend their lives in the canopies of large trees. Katydid females have large spatula or sword-shaped ovipositors with which they glue eggs to the sides of branches. These eggs hatch in the spring, and the young katydids mature in about two months.

Katydids are noted for their style of communication. To attract a mate, male katydids scrape the dorsal parts of their first pair of wings together. Like a bow on a string, this stridulation produces a species-specific song that females hear through ears on their front legs. Female katydids prefer males with the loudest song, because song volume is correlated with body size and body size is correlated with genetic quality. In other words, females don't mate with just any male. They seek out the largest, most powerful male in order to produce the best offspring. Anyone who has spent a summer among large trees is familiar with the different songs of the various species of katydids. In fact, katydids get their onomatopoeic name from their familiar summertime songs. One of the things I have enjoyed as we restore the forest that was once on our property is watching the katydid population return. Each summer since we started planting trees, their nocturnal chorus has grown in volume and diversity. It is just another way I have been surprised by how fast we have been able to make a difference.

A

C

B

With its leaflike wings, *Pterophylla camellifolia*, a common katydid in deciduous forests (A), easily hides amid the foliage of large trees. Katydids glue their eggs to the sides of twigs in the fall (B), where they stay until they hatch in late spring. Nymphs (C) are gangly versions of the adults until their wings develop at maturity.

Not all Tettigoniidae live in trees and not all species have wings. Members of the genus *Atlanticus* spend most of their life on the ground eating low-growing vegetation.

Tree and bush crickets

Order Orthoptera, family Gryllidae, subfamilies Oecanthinae and Eneopterinae
Number of species north of Mexico: 28
Life stages: egg, nymph, adult

Whereas most crickets are dark-colored ground dwellers that lay their eggs in soil and eat whatever they can find, tree and bush crickets are green and spend their lives in vegetation. They will eat insect eggs if they come across them, but most of their diet consists of leaves. The ovipositor of tree and bush crickets is needlelike and is used to insert eggs into the stems of plants like goldenrod and blackberries. These eggs stay in the stems all winter and hatch in the spring. This is one reason it is important not to mow field vegetation in the fall: fall mowing destroys the following year's population of crickets and their role in the food web.

Like katydids, male tree crickets in the genus *Oecanthus* attempt to lure females to them by making chirping songs with their wings. The loudest

Oecanthus latipennis sings within a cupped leaf to amplify its song. Photo by D. H. Funk.

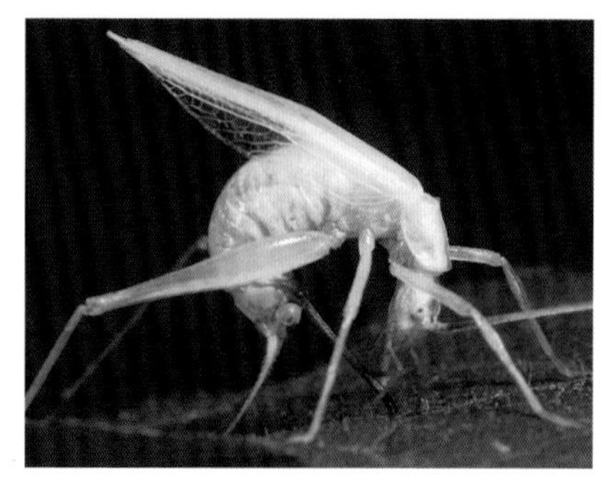

After attracting a female through song, a male tree cricket (*Oecanthus latipennis*) draws a female into position for sperm transfer, lifting his wings and offering her nuptial food from glands under his wings (left). Note the spermatophore dangling from the female's posterior. The male must keep the female occupied during sperm transfer because as soon as his glands run dry, she plucks his spermatophore from her genital opening and eats it (right). Photos by D. H. Funk.

male attracts the most females, so males often cheat a bit by positioning themselves within a cup-shaped leaf that amplifies the song beyond what the male can make without acoustical help. Each male chews a hole in the center of his cupped leaf that is just large enough to accommodate his raised wings during chirping. This ensures that the sound projects directly from the center of the parabolic leaf for maximum amplification.

Once a female selects a male to her liking, a fascinating mating ritual ensues (Funk 1989). To entice the female into the position required for sperm transfer, the male exudes a tasty treat from glands on the top of his thorax. The female climbs on the male's back to eat from these glands, which, not coincidentally, is the ideal position for enabling the male to insert his sperm package (a spermatophore) into her vaginal tract. It is no surprise that the male's goal is to inseminate the female. One might think that would be the female's goal as well, but this isn't the case. Sperm are easy for the female to come by. What she really wants is the nutritional package that accompanies the male's sperm. The female uses these nutrients to make eggs, so as soon as a male inserts the neck of his large spermatophore into her vaginal tract, she can reach around, pluck it out, and eat it! This is not what the male has in mind, as it were, so he has countered the female's hunger for his spermatophore by the evolution of his thoracic nuptial glands. The male produces his tasty secretions just long enough to distract the female from eating his spermatophore before the sperm have completed their transfer. If a male is not strong enough to generate enough thoracic secretions, his sperm end up in the female's stomach rather than in her eggs.

Walkingsticks

Order Phasmatodea, family Heteronemeidae
Number of species north of Mexico: 33
Life stages: egg, nymph, adult

Walkingsticks are aptly named because they look just like perambulating sticks. Their name derives from the Greek word *phasm*, which means "phantom," and describes their cryptic appearance. Not only do they look like sticks, they behave like sticks as well. Walkingsticks spend most of their time absolutely motionless and are therefore extremely hard to spot. They are a large and diverse group in the tropics, but in North America there are only a few common species in four different families. Walkingsticks can either have wings as adults or be wingless, depending upon the species. In some species the females have no wings, while the males are capable of strong flight. Some tropical species are among the largest insects on earth, measuring 12 inches long at maturity. The species most commonly encountered east of the Mississippi, *Diapheromera femorata*, is about four inches long.

Walkingsticks live primarily in the canopy of mature forests, where they eat leaves throughout their entire life. Females are not very sophisticated when it comes to laying their eggs. They simply pop them out the back one at a time and drop them to the forest floor. Most eggs hatch the following spring, but some remain dormant until a second year has passed. The young walkingsticks start off eating perennials in the understory, but soon they climb the nearest tree, and there they stay until the leaves drop. When I lived in a housing development with many large oaks, I would often encounter walkingsticks on the side of my house after leaf fall.

Walkingsticks are the picture of their name. A *Diapheromera femorata* nymph blends right in with its clematis support.

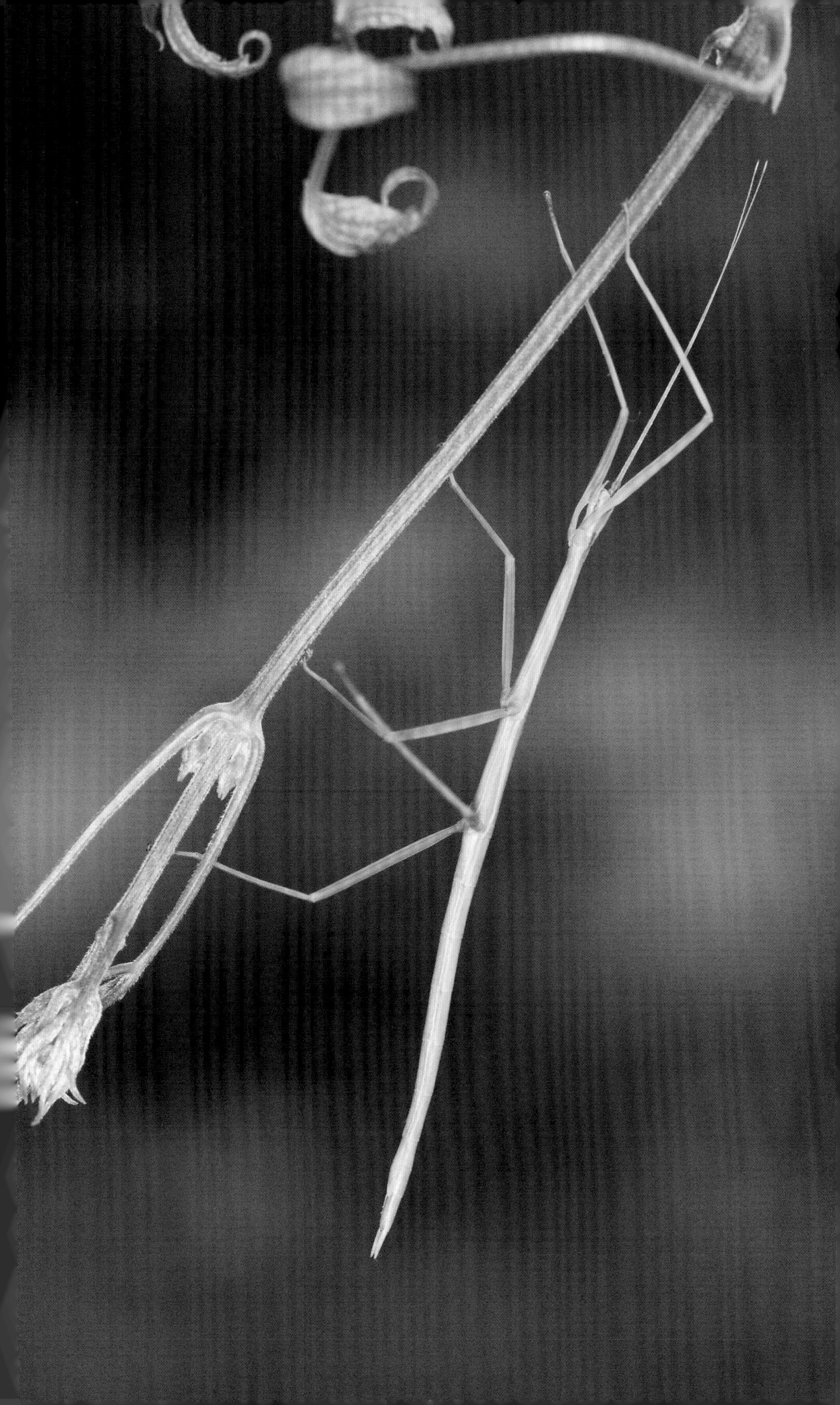

Plant bugs

Order Hemiptera, family Miridae
Number of species north of Mexico: 1750
Life stages: egg, nymph, adult

Plant bugs are the first of several groups of insect herbivores mentioned here that suck plant juices through mouthparts specialized to pierce plant tissues. For plant bugs, these tissues are the parenchymal cells that constitute the layer of a leaf between the upper and lower epidermis. Often plant bugs can suck from leaves and leave little visible damage. Other times, the leaf reacts to the bug's saliva and becomes discolored at feeding sites. There are many species of plant bugs, and they occur on most types of native plants, although they are most common in field vegetation. The females inject their eggs into soft plant stems with a sharp ovipositor that folds up into a slit along the ventral (underside) surface of the abdomen when not in use. Within their stem shelter, the eggs are protected from predators and desiccation. Plant bugs are relatively small, and most are rather drab. The gardener may never notice them. Nevertheless, their large numbers make plant bugs an important component of the food web.

Plant bugs (Miridae) may not be the stars of the garden, but their dull coloring does not prevent them from being snapped up by larger insects.

Lace bugs

Order Hemiptera, family Tingidae
Number of species north of Mexico: 140
Life stages: egg, nymph, adult

Unfortunately, most gardeners come to know lace bugs through unpleasant encounters with the azalea lace bug, an Asian species that was introduced with Asian azaleas in the early 1900s. Native species are better behaved, however, and are rarely seen. These tiny bugs, like plant bugs, live off the chlorophyll in leaf cells. They always eat from the undersides of leaves and so are easy to miss. Most lace bug species are excellent examples of herbivores that specialize on one or only a few closely related plant species. For example, the cherry lace bug develops only on cherry trees; the sycamore lace bug can eat nothing but sycamores; the ash lace bug nothing but ashes; the alder lace bug is confined to alders. Morphologically, they are unique among insects in having areolate thorax and wings. That is, these structures are filled with small spaces that make them look like camouflage netting. The effect of this design is to break up the body outline and help the lace bugs blend into their leafy backgrounds. Most species of lace bugs have multiple generations per summer and so are excellent sources of insect biomass for other animals.

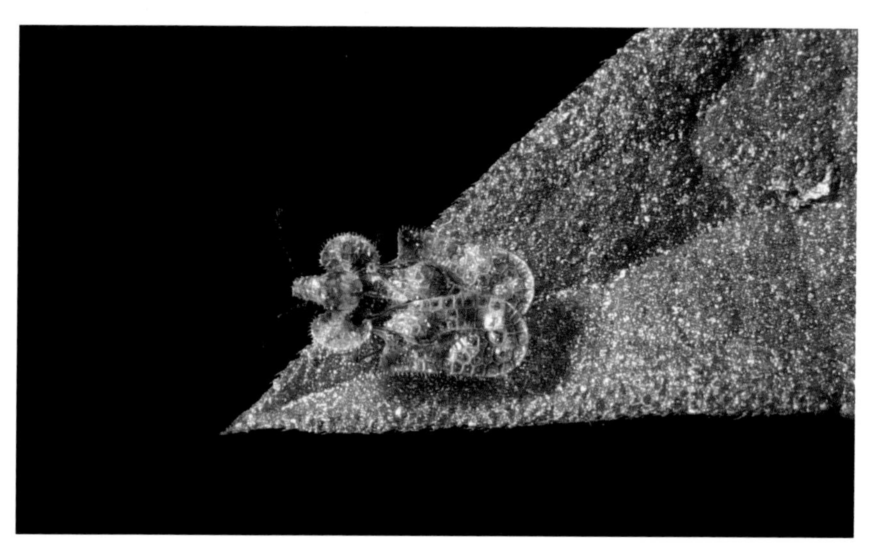

Corythucha marmorata, a lace bug common on goldenrod, spends most of its life on the underside of leaves.

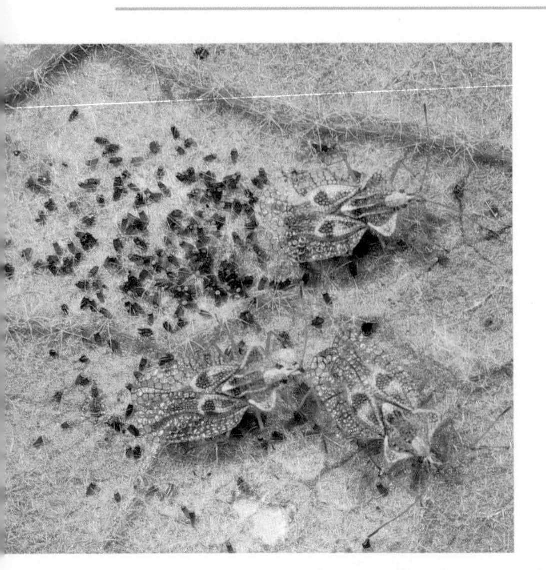

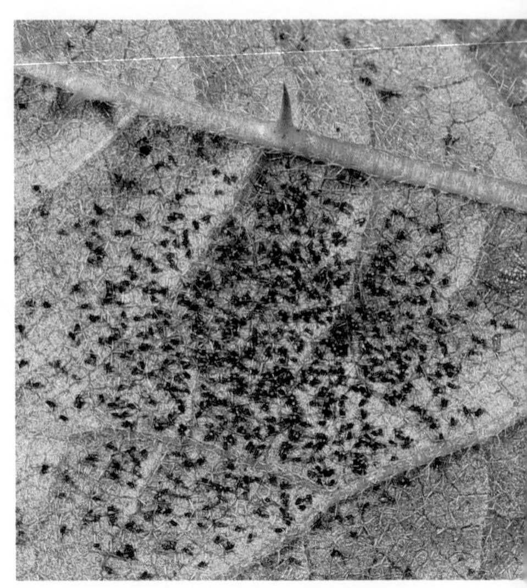

As soon as one *Gargaphia solani* female starts to lay her eggs, which look like small black dots on the underside of the leaf, other gravid females are attracted to the site and add their eggs to the lot (left). After all the eggs are laid, only the female who started the egg mass remains to guard it (right).

A few lace bug species take care of their eggs and the resulting nymphs until they mature. *Gargaphia solani* mothers stand near their eggs and subsequent young bugs and fan their wings at any approaching predator. Without any real weapon to use for defense, all they can do is harass predators that find their young. Being annoying appears to be good enough, though, because often the predator will turn and hunt elsewhere. Though necessary, all of this guarding takes lots of time—half of the mother's adult life span for a single clutch of eggs. While she is guarding one brood, a mother cannot lay any additional eggs. So *Gargaphia* has developed a reproductive alternative that females use whenever they have the opportunity. If a female who is ready to lay eggs comes across another female *Gargaphia* who is already guarding her eggs, the first female lays her eggs in the egg mass of the guarding mother—and then leaves! This is a good deal for the "egg dumper" because she can now go off and lay more eggs while her first clutch is being cared for. It's like taking the kids to the babysitter but never returning to pick them up. But don't feel bad for the guarding female who accepted the dumper's eggs. She now has a buffer of unrelated eggs surrounding her own eggs. When a predator succeeds in eating some eggs, more often than not it is the egg dumper's eggs that are lost (Tallamy & Horton 1990).

Seed bugs and their relatives

Order Hemiptera, family Lygaeidae
Number of species north of Mexico: 250
Life stages: egg, nymph, adult

It's no surprise that seed bugs feed primarily on seeds, usually those that have already fallen from the plant. They have sucking mouthparts that they insert into the seed—a tough plant part to imbibe as if it were a milkshake. Seed bugs have solved this dilemma by digesting the tough seed *before* they suck it up. They inject saliva into the seed and wait for the proteolitic enzymes in the saliva to dissolve the seed's innards. Once everything inside is mush, the bug sucks up the whole mess. As you might expect, the more saliva that is pumped into the seed, the faster and more completely the seed is dissolved. For this reason, seed bugs often feed gregariously, with all members of the group "spitting" into the seed and sucking up the end product together. This eating strategy allows seed bugs to eat seeds that would be too large for a single bug to tackle.

Because most seed bugs eat seeds that are on the ground, these insects are not usually encountered by gardeners. However, two species eat milkweed seeds before they dehisce from the seedpod. *Oncopeltus fasciatus* and *Lygaeus kalmii*, the large and small milkweed bugs, are consequently up on the plant and plainly visible, particularly after seeds have started to develop in late summer. Like other insects that eat the different species of milkweeds, milkweed bugs are black and red, a universal coloration that serves as a warning to vertebrate predators, advertising the fact that these insects taste terrible. Milkweed plants produce cardiac glycosides, secondary metabolic compounds that can cause heart attacks in vertebrates. They pose no problems for insect specialists on milkweed, though, which have evolved metabolic adaptations for breaking this toxin down.

The observant gardener will notice that the large milkweed bug does not appear until later in the summer season, and that when it first shows up, it is always as an adult. Where has it been hiding out? In the late 1960s, Hugh Dingle, then at the University of Iowa, discovered that *Oncopeltus fasciatus* migrates all the way from its overwintering grounds along the Gulf Coast to as far north as Canada (Dingle 1972). That's quite a trip for an insect just over a half-inch long, but this species uses migration to solve the nutritional problems that come from specializing on the seeds of one particular group of plants. In the spring and early summer, milkweed plants in the North have not yet started to produce seedpods. *Oncopeltus* would have nothing to eat for months if it spent the winter in northern regions and then emerged in the spring looking for food. However, by flying south to the Gulf States, milkweed bugs can find a steady supply of milkweed seeds, which are produced all year long in the South. Then, as milkweed plants start to mature from south to north, the bugs just follow their supply of fresh seeds.

Adult (above) and nymphs (right) of *Oncopeltus fasciatus* suck the juices from milkweed seeds while the seeds are still in the pod.

Box elder bugs

Order Hemiptera, family Rhopalidae
Number of species north of Mexico: 37
Life stages: egg, nymph, adult

Larger and darker than its close relatives, the box elder bug (*Boisea trivittata*) is a morphological oddity in its family, Rhopalidae. It nevertheless is a very common insect wherever mature box elder trees occur. As black and red seed-sucking specialists, box elder bugs look and act much like milkweed bugs. You will never confuse them, though, if you remember that specialists are always associated with their host plants. Milkweed bugs are always around milkweed, and box elder bugs are found exclusively on or near box elder—and not just any box elder. Because these insects primarily eat seedpods, they favor the female trees, which routinely produce copious amounts of seed. Females lay their eggs in the spring near piles of box elder seeds from the previous fall. As the nymphs are developing on old seeds that have fallen to the ground, a new batch of seed is produced by the tree and held for months on the branches. The bugs eventually climb the tree and cluster on the seed clumps, eating them until they complete their development to adulthood.

Both the adult (right) and nymphs (above) of *Boisea trivittata* rely on box elder seeds to complete their development.

Most people encounter box elder bugs in the fall, when cool weather sets in. Adults of the species overwinter in protected places, sometimes emerging to bask in the sun on warm days. People who have female box elder trees near their houses often find hundreds of these insects warming themselves on south-facing walls or, even more disconcerting, moving inside for the winter. They are entirely innocuous, but these little guys can get in the way when they take up winter residence in your home.

Leaf-footed bugs

Order Hemiptera, family Coreidae
Number of species north of Mexico: 80
Life stages: egg, nymph, adult

Leaf-footed bugs are so named because the tibia of the hind legs is often (but not always) flattened into a broad, leaflike structure. In some tropical species, these structures are often quite large and colorful. In fact, most species of leaf-footed bugs are rather large and robust insects. They can resemble common predators called assassin bugs, but all leaf-footed bugs are herbivores that suck the contents of leaf cells. Though most North American species are gray as adults, the females of many species produce brilliant golden eggs, which they lay in a row on vegetation. I have never seen more beautiful insect eggs.

Acanthocephala terminalis, a common leaf-footed bug in the East, gathers on ash trunks in the spring to mate.

Shield bugs

Order Hemiptera, family Pentatomidae
Number of species north of Mexico: 200 plus
Life stages: egg, nymph, adult

Shield bugs are insects of moderate size that look very much like the shields used by medieval knights of old. One subfamily (Asopinae) contains predatory species, but all of the other shield bugs are herbivores. Like the other phytophagous hemipterans we have discussed, these bugs suck the juices from leaf cells rather than from the plant's vascular system. They are common insects, but under natural conditions they never undergo eruptive population growth. Shield bugs lay huge, barrel-shaped eggs that hatch into flattened round nymphs. Because plants are often difficult to digest, most insect herbivores require the help of symbiotic protozoans and bacteria that are passed from the mother to the young through the egg. Shield bug hatchlings must suck these up from a pool left in the eggshell. It is not unusual to find a cluster of first-instar shield bugs—the first of five developmental stages—huddled around their empty eggshells as they imbibe the mixture that will enable them to eat plant material the rest of their lives.

A shield bug from Colombia stands guard over her nymphs.

Shield bugs (Pentatomidae) are aptly named, resembling medieval shields (left). These insects are also called stinkbugs because of the defensive odor they produce from glands on their thorax (right).

Shield bug eggs are large and barrel-shaped (left). After hatching, the nymphs take their first meal from the fluids remaining in their eggshells (right).

Females of many species of shield bugs, along with the members of several closely related families, protect their eggs and young nymphs from predators and parasitoids. After laying a clutch of eggs, a female will stand directly over them to shield them from harm. If a predator approaches from the left, she tilts her body left. This investment by the mother is very effective as long as she doesn't lay more eggs than she can cover with her body. Any female who lays more eggs than she can cover loses those extra eggs to predators. Thus, the genes promoting larger clutches are constantly weeded out of the population. Over time, this selection pressure has molded clutch size in species with maternal care to conform very nicely to the female's body size.

Cicadas

Order Hemiptera, Family Cicadidae
Number of species north of Mexico: 157
Life stages: egg, nymph, adult

Cicadas are a group of large insects that are poorly understood, even by entomologists who have studied them for years. Some people mistakenly call them 17-year locusts, but they are not locusts at all. Locusts are grasshoppers, with chewing mouthparts. Cicadas have piercing-sucking mouthparts, which they use to suck fluids from the xylem of a tree's vascular system. They are noted for the males' loud singsong buzz, which is used to attract females for mating. We know so little about cicadas because they spend almost all of their lengthy life cycle underground, sucking on tree roots. After years in the ground, they finally crawl to the surface and molt to adults. We often see a final nymphal skin fastened to a tree trunk by the claws of the front legs, the slit down the back showing where the adult emerged.

In eastern North America, cicadas fall into two distinct groups: the periodical cicadas and the annual cicadas. In one of the most fantastic feats of biological synchronization known, all of the individuals of three species of

Usually, the only physical evidence we see for the presence of cicadas is the molted nymphal skin left clinging to a tree trunk.

Periodical cicadas can emerge in great numbers. Here my daughter plays with individuals from the 1987 brood that emerged in Newark, Delaware.

Annual cicadas are easy to distinguish from periodical cicadas. Annual cicadas (bottom) are larger and are trimmed in green. Periodical cicadas (top) are about ¾ the size of an annual cicada and sport orange markings rather than green.

Cicadas insert their eggs into woody tissue with a strong ovipositor (above). This action often kills the branch terminal, causing a characteristic "flagging" (below).

periodical cicadas time their molt to adulthood to within a few days of each other, after spending 17 years as nymphs below ground. This event occurs in June and results in the sudden appearance of millions of orange and black cicadas in our woods and yards. The largest numbers are seen in woodlots that have been undisturbed for the last 17 years. Farther south there are four additional species of periodical cicadas that emerge every 13 years.

Annual cicadas are slightly larger than periodical species; they are etched in green rather than orange; and they emerge in midsummer rather than late spring. They differ from periodical cicadas in two other important ways as well: the individuals of any given population are not synchronized in their emergence, and their populations are far smaller than the periodical broods. Although researchers suspect that most annual cicadas develop much faster than periodical species, we are not sure how long the majority of species live underground as nymphs. We do know that some individuals emerge every year, which is why we call them annual cicadas, even though they have been underground many years.

Both periodical and annual cicadas use strong ovipositors to bury their eggs in the soft woody tissue of young tree twigs. Often the tissue damage caused by egg-laying kills the terminal portion of the twig. The twig then bends down at right angles and looks like a small flag. Large populations of periodical cicadas can prune significant portions of new growth from trees every 17 years. The eggs hatch after a few weeks and the young nymphs fall to the ground, where they immediately tunnel under the soil in search of a tree root. It is assumed that once they find the vascular system of a root, they move very little for the remainder of their long development.

One of the great challenges for evolutionary biologists has been to explain the curious life cycle of periodical cicadas. Why synchronize emergence, and why wait 17 years to do it? The most popular explanation is the need to avoid predators and parasites. If all individuals in a brood emerge as adults at the same time, there will be too many cicadas for birds, squirrels, and chipmunks to eat; and most individuals will live long enough to reproduce. Predators cannot adjust their own populations to take advantage of the sudden influx of food from an emerging brood because the time interval between broods (13 or 17 years) is too long. This is called the "predator satiation" hypothesis; it only works when populations are large enough to swamp the predators that are around every year. Unfortunately, this is no longer the case in many places. The last cicada emergence near our house was in 2004. It was smaller than in the past, and I watched a single squirrel catch and eat every cicada (dozens of them) that emerged from the roots of a large willow oak. Periodical cicadas are declining in many parts of their range, almost certainly because of the pesticides used by lawn care companies, and because of sprawl itself. When our woodlots are leveled for housing developments, the nymphs below ground starve by the millions.

Treehoppers

Order Hemiptera, family Membracidae
Number of species north of Mexico: 258
Life stages: egg, nymphs, adults

Although treehoppers are relatives of cicadas, the two do not look much alike. Treehoppers are quite a bit smaller than most cicadas and are distinguished from all other insects in having the upper plate of their exoskeleton, the pronotum, expanded into bizarre shapes. Many of the species common in North America have a prothorax that looks like a rose thorn, but in other parts of the world it can look like lichens, a string of beads, or an umbrella. No one yet knows what has caused the evolution of these shapes. When the pronotum is a tricornered spike, it makes the treehopper difficult for predators to swallow. But why, then, don't all species have a spiked pronotum? Other ideas suggest a sensory function associated with pronotum shape, but this is poorly understood.

Treehoppers are so named because many species are arboreal and, when disturbed, the adults jump. Maternal care of eggs and nymphs is common in this family of insects. Scientists such as Rex Cocroft of the University of Missouri have discovered that mothers and offspring communicate with each other by generating vibrations that travel through the branch they are sitting on and are detected through the insects' feet (tarsae). For example, in

Ceresa diceros (left) is a common treehopper in the East that feeds primarily on oaks. Although most membracid species live in trees, many like *Entylia bactriana* (right) develop on herbaceous plants.

This treehopper from Costa Rica guards eggs she has buried in the midrib of a leaf. She herself is protected from predators by the sharp horns protruding from her prothorax.

the tropical species *Umbonia crassicornis*, nymphs perched on thin branches of leguminous trees feed in groups on tree sap while their mother stands guard just below them, ready to intercept any predator that might walk up the branch (Cocroft 1999). This defense fails when a predator attacks from the air, as do sphecid and vespid wasps. These wasps hover over the nymphs before grabbing one off the branch. Wasps are very effective predators, and they take a heavy toll on the young treehoppers. Through branch-borne vibrations, nymphs can alert their mother that a wasp is hovering nearby. The entire group hums an alarm song, a thrum-thrum-thrumming sound, which the mother "hears" through her feet and immediately responds to. She moves to the center of the nymphs and attempts to kick the wasp out of the sky. This sounds like a feeble defense, but her kick is powerful enough to hurt (if she happens to kick you instead); and if she connects with the wasp, the impact sends it tumbling.

Leafhoppers

Order Hemiptera, family Cicadellidae
Number of species north of Mexico: 2500
Life stages: egg, nymph, adult

Even smaller relatives of cicadas are the large group of species commonly called leafhoppers. When magnified, these tiny insects look like speedboats, and many are brilliantly colored. Because of their small size, a single plant can host hundreds or even thousands of leafhoppers without any noticeable effects. Although many are quite host-specific, leafhoppers as a group develop on a wide variety of plants, from annual forbs to bushy perennials and large trees. They are very numerous in old fields and are an important component of the diet of swallows and other small birds. They typify hundreds of types of small insects that go unnoticed by the suburban gardener.

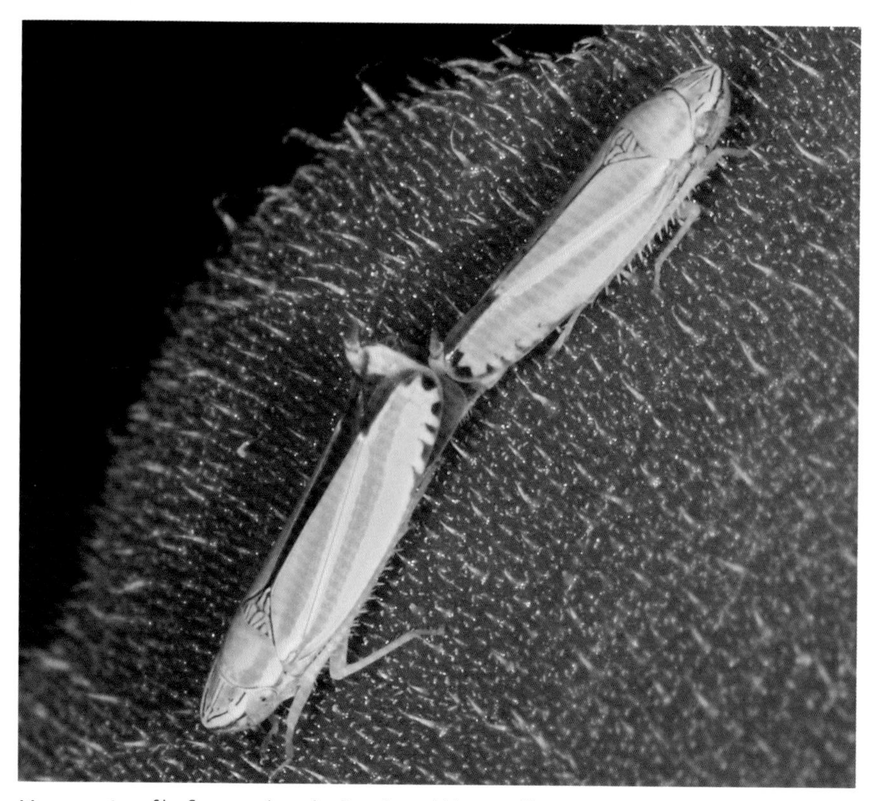

Many species of leafhoppers have bodies shaped like speedboats. *Graphocephala versuta* is one of the more beautiful species in North America.

Froghoppers and spittlebugs

Order Hemiptera, family Cercopidae
Number of species north of Mexico: 54
Life stages: egg, nymph, adult

If you have ever walked through a field in June while wearing shorts, you already know what a spittlebug is. Attached to stems (goldenrod is a favorite) about two feet off the ground are white masses of bubbles that look just like spittle. Perhaps knowing that the bubbles are just plant sap might make it less disconcerting when the "spittle" smears across your bare leg. What you do not see is the delicate froghopper nymph inside the mass of bubbles. The nymph, or spittlebug, has created the bubble mass as a defense against predators. Predators don't like to get gooey any more than we do. The spittlebug is able to create its little bubble house by sucking more plant sap than it can digest and passing the extra through its digestive system and out its anus, almost unchanged from what it was in the plant. Just before the sap leaves its anus, the spittlebug pumps a tiny amount of air into the sap, creating a bubble that lasts for days. As the nymph makes more and more sap bubbles, these well up around its soft body and eventually cover it completely. Spittlebugs complete their development inside their bubble houses and don't come out of hiding until they have matured. As adults, they look like stocky leafhoppers and are just as jumpy.

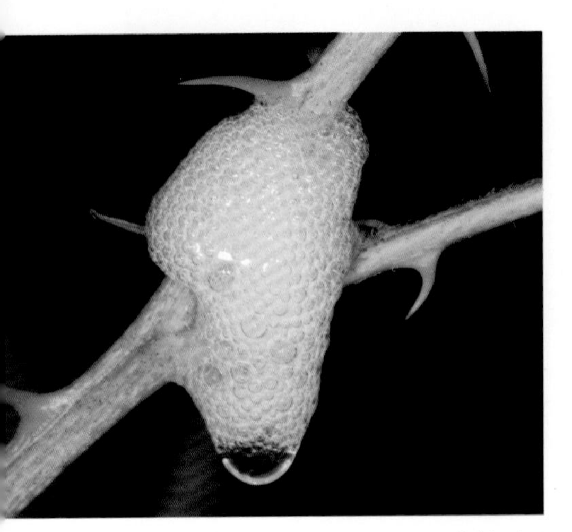

The "spittle" so common in country fields during early summer is actually the home of spittlebug nymphs (left). Adult spittlebugs, like *Philaenus leucophthalmus*, are called froghoppers (right).

Aphids

Order Hemiptera, families Aphididae, Adelgidae, Eriosomatidae, Phylloxeridae
Number of species north of Mexico: 1450
Life stages; egg, nymph, adult (often parthenogenic)

Throughout the world, simply because they are so common and so numerous, tiny aphids play an enormously important role in transferring the sun's energy from plants to larger animals. Anthony Frederick George Dixon, the famed British entomologist who dedicated his life to the study of aphids, once estimated that a single acre can support 2 billion aphids on vegetation and another 260 million below ground on plant roots (Dixon 1997). Aphids are such a constant and dependable feature of most habitats that entire insect families have become their specialized predators. Three good examples of taxa that have specialized on eating aphids are coccinellid ladybird beetles, syrphid flower flies, and chrysopid lacewings.

Because aphids usually reproduce parthenogenically and viviparously, that is, females lay young nymphs, skipping both mating and the egg stage, populations can build to high numbers very quickly.

Aphids become so numerous each summer for two reasons. First, they use plant resources efficiently. One acre of alfalfa can indefinitely support aphids at a biomass equivalent to that of an elephant, whereas an elephant would eat all of the alfalfa in that acre in a matter of hours (Dixon 1997). Aphids usually exploit plants with few measurable effects on those plants, particularly when the plants are not water stressed. Each tiny aphid inserts its sucking mouthparts into the plant's phloem tubes. Phloem is extremely low in nitrogen, and aphids have to pass tremendous quantities of phloem through their digestive tracts to extract enough nitrogen for their growth and reproduction. Over the eons, aphids have developed a specialized digestive device called a filter chamber, which removes the small amount of nitrogen from the ingested phloem and shunts most of the remaining sugar water directly out the anus. We call this end product "honeydew," which, because of its mildly sugary content, is attractive to numerous types of insects, particularly ants. As any picnicker knows, ants love sugar, and they are willing to protect a group of aphids in order to enjoy exclusive access to their honeydew. It is difficult to find an aphid aggregation that is not being tended like cattle by a group of ants.

The second reason aphids are so good at making more aphids is that their complex life cycle works beautifully for getting the most nitrogen from plants throughout the year. There are innumerable variations, but if there were an "average" aphid life cycle, it would be something like this: Most species spend the winter as an egg laid on a woody host, typically a deciduous tree. In the spring, as the host tree's leaves flush and grow rapidly, the egg hatches into a "stem mother," the individual responsible for creating a large clone of aphids later in the season. The stem mother grows to maturity on the nitrogen-rich young leaves of the host tree and then starts giving birth to nymphs. Notice that she is able to produce young without mating with a male (parthenogenesis) and that her offspring are born live (viviparity). This simplifies—and speeds up—reproduction a great deal. At first, all of the stem mother's offspring are wingless females that begin to make babies themselves. After the host plant's leaves have fully expanded and the spring growth spurt is over, the nitrogen level in the phloem drops, and it behooves the aphids to find a more nutritious host. So the next generation of aphids produced by the stem mother and all of her offspring are winged females, which take flight in search of their summer host. This summer host is usually a herbaceous perennial that is undergoing rapid growth and sporting relatively high nitrogen levels in its phloem. Although millions of aphids

Carpenter ants protect a group of aphids on *Viburnum dentatum* in exchange for access to honeydew.

become meals for bats and dragonflies during their migration to the summer host, some arrive successfully and start to reproduce—again through viviparity—several more generations of wingless females. As the summer wanes, the nitrogen levels in the summer host's phloem also drop, sending the signal to produce for the first time a sexual generation of aphids: both males and females with wings. When these nymphs mature, they fly from the summer host back to the winter host, which is now transporting the nitrogen from its leaves back to the roots for winter storage. The aphids mate, and the females take advantage of the nitrogen that is being moved by the tree to produce the eggs that they will lay on the woody host, to start the cycle all over again the following spring.

Leaf beetles

Order Coleoptera, family Chrysomelidae
Number of species north of Mexico: 1827
Life stages: egg, larva, pupa, adult

There is much to say about beetles that eat plants, because there is so much to say about beetles in general. If diversity is a measure of success, beetles are by far the most successful group of multicellular organisms alive today. There are already well over 300,000 species of beetles that have been described by taxonomists, and there may well be that many unnamed species remaining to be described. To put these numbers in perspective, there are 6 times as many described beetles as there are all vertebrates combined, and 34 times more beetle species than bird species. In fact, 30 percent of all animals are beetles.

One of the reasons beetles are so numerous is that they are very good at eating plants. Every terrestrial plant on earth has at least one species of beetle that can eat it, and many plants support different beetle species on their leaves, their shoots, their roots, their flowers, and their pollen. It should not be surprising, then, that one of the largest families of beetles is also one in which all species are herbivores. The Chrysomelidae, whose members

The family Chrysomelidae includes some of our most beautiful beetles. The dogbane beetle (*Chrysochus auratus*) is one of my favorites.

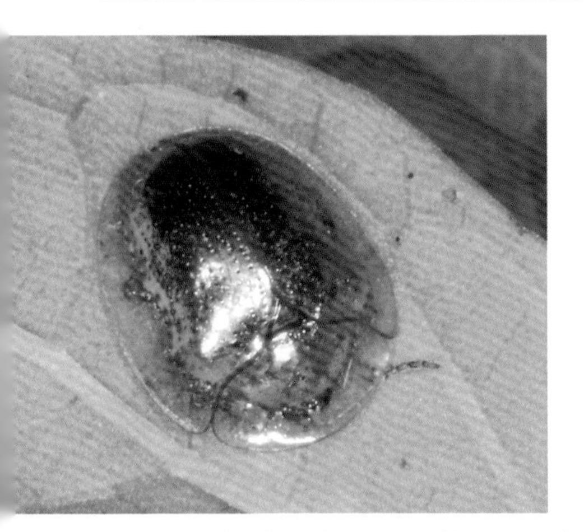

The tortoise beetle, with its covered head, is aptly named for its appearance (left). Larvae of many tortoise beetles protect themselves from predators by carrying an umbrella of their feces arched over their bodies. *Cassida rubiginosa*, a common species on thistle, uses this tactic to reduce its appeal (right).

are known as the leaf beetles, include some of the most beautiful, brightly colored insects in the world—as well as some species like Colorado potato beetle and western corn rootworm that we could do without. Because different groups of leaf beetles specialize on different plant parts, several groups have become quite distinctive in their looks and habits. For example, flea beetles (subfamily Alticinae) are notable for enlarged hind legs that allow them to escape predators by jumping, while tortoise beetles (subfamily Cassidinae) have expanded pronotums that usually cover their heads. In their larval stage, some leaf beetles are called leaf miners (subfamily Hispinae) because they eat the parenchymal cells packed between the upper and lower surface of leaves, and others are called rootworms (subfamily Galerucinae, tribe Luperini) because they develop exclusively on plant roots.

Leaf beetles defend themselves against predators in ingenious ways. Tortoise beetle larvae, for example, have two small hooks on their rear ends that they use to carry an umbrella constructed of their feces. Predators are reluctant to attack a larva shielded in excrement, so this defense is quite effective. Other groups of leaf beetles rely on chemical deterrents to repel predators. The larvae of species such as the cottonwood leaf beetle (*Chrysomela scripta*) have a row of glands on each side of their bodies. When a predator approaches, the larva produces from each gland a drop of highly volatile liquid, which smells terrible. To make this defense more effective, larvae usually stay in

Many chrysomelids are host plant specialists. For example, *Labidomera clivicollis*, the milkweed leaf beetle, eats only milkweed.

To ward off predators, larvae of *Chrysomela scripta* secrete foul-smelling chemicals in a mass of white bubbles from glands on their abdomen.

groups. If one beetle exudes its defensive droplets, they all do, saturating the air around the entire group with the repellant. As soon as the danger is gone, all of the larvae reabsorb the defensive droplets back into their glands for use another day.

Long-horned beetles

Order Coleoptera, family Cerambycidae
Number of species north of Mexico: 900
Life stages: egg, larva, pupa, adult

Another large family of plant-eating beetles is the Cerambycidae. Close relatives of leaf beetles, cerambycids, or long-horned beetles, can be recognized by their long antennae. Many species sport antennae that are longer than their entire bodies. Long-horned beetles also differ from leaf beetles in the way they exploit plants. Long-horned beetles are borers as larvae (called round-headed borers) and develop in the heartwood of trees or the pith of nonwoody stems. Woody tissues are one of the poorest of all food resources and are extremely low in the nitrogen needed to build protein. What's more, the nitrogen that does occur in wood is tied up in cellulose that long-horned beetles cannot directly digest. Like termites, long-horned beetles have several species of microorganisms that live within their guts. These microorganisms digest the cellulose eaten by the beetle larvae and release the nitrogen locked up in the cellulose. The microorganisms also contribute to the meager amounts of nitrogen in cellulose with the nitrogen within their own waste

Most cerambycids sport long antennae.

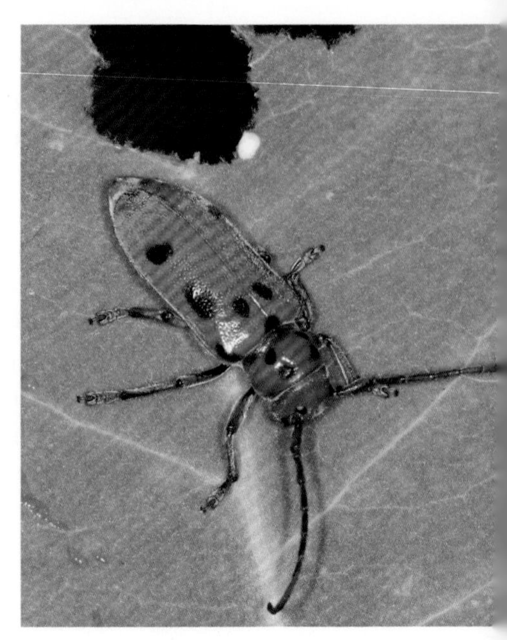

The red milkweed beetle (*Tetraopes tertraophthalmus*) blocks the flow of latex sap in milkweed leaves by snipping the midrib near the terminal of the leaf (left), enabling the beetle to eat the tip of the leaf without encountering the toxic substance (right). Note the white marks on the leaf where the beetle has snipped through the midrib.

products. Despite these complex symbiotic relationships, it still takes some long-horned species incredibly long periods of time to reach maturity. The record is held by a long-horned beetle that emerged from a piece of furniture 40 years after the furniture was carved from a tree which presumably contained a young larva (Jaques 1918).

One of the cerambycids most frequently encountered by homeowners is *Tetraopes tertraophthalmus*, the red milkweed beetle. This species develops as a larva within the stems of milkweed plants and is one of the few insects that have evolved a strategy for eating milkweed leaves as an adult. Milkweed leaves are laced with canals that are filled with a distasteful latex sap that gums up insect mouthparts when exposed to air. Most insects avoid milkweed tissues because they cannot find a way to eat them without taking a bath in the milky sap. The red milkweed beetle, however, disables the latex defense system of milkweeds by snipping the midrib of the leaf in two or three places about two inches down from the tip of the leaf. This blocks the flow of latex to the canals in all of the leaf tissues outside the snips. The beetle is then free to eat the leaf terminal without exposure to the latex.

Metallic wood-boring beetles

Order Coleoptera, family Buprestidae
Number of species north of Mexico: 762
Life stages: egg, larva, pupa, adult

As adults, metallic wood-boring beetles look much like bullets. They are often coppery in color, and their flattened heads taper to pointed rear ends. Like long-horned beetles, metallic wood-boring beetles develop as larvae in the heartwood of trees. The larvae are called flat-headed borers and are commonly encountered along with cerambycid round-headed borers by people who split wood for wood-burning stoves. Though they may take several years to reach maturity, adults typically do not live more than a few weeks.

Because wood-boring beetles have better survival rates in trees that are not healthy enough to expel them with sap as they try to eat their way through the bark, these insects are very good at finding trees that are stressed or have been weakened by disease. One species in the West has developed specialized heat sensors that can detect infrared radiation emitted by burning trees up to 80 miles away. These beetles fly to the top of mountains and scan the horizon in search of forest fires. If they detect a fire, they fly to it and lay their eggs in the weakened trees even before the trees stop burning.

Most species of Buprestidae are bullet-shaped and have a metallic sheen.

Stag beetles

Order Coleoptera, family Lucanidae
Number of species north of Mexico: 24
Life stages: egg, larva, pupa, adult

Stag beetles are yet another group of insects that develop in wood, but they differ from long-horned beetles and metallic wood-boring beetles in that they only use trees that are entirely dead. The insect's common name derives from the allometric growth of the male's head and mandibles; both are much larger than they should be, given the size of the beetle's body. In some species, the mandibles are actually branched and resemble the antlers of deer or elk. Mandibles on females, in contrast, develop in proportion to their body

Stag beetles like *Pseudolucanus capreolus* are declining throughout their range as the large dead trees they require for reproduction disappear.

size and thus are diminutive compared to the males'. The stately antlerlike mandibles, huge head, and mahogany color of male stag beetles make them one of our most striking insects.

Why should male mandibles be so much larger than the mandibles of females? It is certainly not because males eat more than females; neither sex eats anything after emergence as adults. Noticeable differences between the males and females of a species usually arise in the context of competitive mating, and stag beetles are an excellent example. Males use their huge mandibles to fight with other males for mating opportunities with receptive females. These beetles usually mate in the canopy of mature forest trees. If two males find a female at the same time, they attempt to lift each other entirely off the branch with their mandibles. The first beetle to succeed in this endeavor hurls his opponent to the ground. As soon as the victorious male disposes of his rival, he mates with the female, who waits passively nearby. By the time the deposed male finds his way back up into the tree, the deed is done. Because the beetle with the largest mandibles is typically the winner in these contests and ends up with the female, more genes coding for large mandibles are passed on to future generations than genes for small mandibles and, *voilà!* you have "sexual selection" for large male mandibles. No such selection acts on females, so their mandibles are unimpressive. By cooperating in this mating game, the female is assured of copulating with a strong male who will pass the genes for large mandibles on to her offspring.

Stag beetles are declining throughout their range because they require large dead trees from mature forests to complete their larval development. Our fragmented young woodlots often do not contain trees that are large enough or dead enough to support populations of these impressive beetles.

Bess beetles

Order Coleoptera, family Passalidae
Number of species north of Mexico: 4
Life stages: egg, larva, pupa, adult

A small but fascinating family, the passalids, or bess beetles, spend all but a few days of their life within fallen trees. These relatively large beetles (topping an inch in length) are unusually uniform in looks across species. Mature adults are jet black; teneral (recently matured) beetles remain an auburn-brown for several weeks before turning black. Although there are only four species of bess beetles in North America, they are relatively common wherever there are mature hardwood forests.

What distinguishes bess beetles from other beetle species is their level of sociality (Tallamy & Wood 1986). Bess beetles live in large logs in extended family groups within a series of interconnecting galleries that they excavate with their mandibles. Each family unit is independent of other families: in fact, they are hostile toward outside bess beetle groups and guard their galleries against intruders. Only the primary parents reproduce within a gallery; the first offspring to mature typically remain with their parents for months and lend a hand in rearing additional brothers and sisters. There is much to be done in this regard, for getting enough nutrition from cellulose to complete development is one of the most difficult ways to make a living in the world of insects. Young larvae are not large enough, nor strong enough, to chew through wood themselves. Moreover, they lack the internal symbionts required to digest cellulose once it is consumed. This problem is tackled by parents and older siblings, who chew up bits of wood for the larvae and mix the shavings with their feces to inoculate them with the microorganisms that the younger larvae need for digestion. When a larva reaches its fifth and final instar, an adult sibling builds a pupal chamber for it out of wood shavings. The larva rests within this chamber until it ecloses to a teneral adult and joins the family workforce. How do family members know what the needs of their relatives are? They talk to each other, of course. Bess beetles, like grasshoppers and katydids, rub body parts against each other and produce audible stridulations that convey specific messages to nearby family members. If you pick up a bess beetle, it will talk to you too. You can guess what it is saying!

Bess beetles like *Odontotaenius disjunctus* are one of the few species of beetles that live in extended family groups. Older adults (C) are jet black and enlist the help of mahogany-colored teneral adults (A) to help rear the next batch of young. Bess beetle larvae eat wood shavings prepared by their parents and older siblings (B).

The only time bess beetles leave their home log, unless it has decayed beyond the point of further use, is when they become reproductively mature and attempt to start their own family in a new, unoccupied log. Every year, this part of the bess beetle's life cycle becomes increasingly dangerous, for there are more bess beetles than there are suitable large logs. Most of our old-growth forests have been logged and replaced either by young trees too small to support a bess beetle family or by no trees at all. Beetles that fail to find a new home die without ever reproducing.

June beetles, chafers, hercules beetles, and their relatives

Order Coleoptera, family Scarabaeidae
Number of species north of Mexico: 628
Life stages: egg, larva, pupa, adult

Members of the large family Scarabaeidae all eat plants in one form or another. Dung beetles, which I will not discuss here, eat plants after their leaves and flowers have passed through the gut of a vertebrate—that is, after the plants have been converted to dung. Most other scarab species develop as larvae on roots, detritus (plant tissues that are no longer on a plant), or rotting wood. Adults typically eat leaves or flower petals. The family contains some of our largest insects by weight. All species have thick exoskeletons and remind me of little tanks. The larvae are white and C-shaped, in classic grub morphology.

Scarabs separate nicely into easily distinguished lineages. The subfamily Melolonthinae contains many species commonly called June beetles (May beetles if you live in the South). These are the familiar brown beetles that bang into your screened windows at night because they are attracted to lights. Rose chafers also belong to this group. They are smaller than June beetles (about half an inch long), have longer legs, and are tan rather than dark brown. Rose chafers can be common in areas with light sandy soils. The larvae of June beetles and rose chafers eat the roots of grasses and herbaceous dicots and periodically are quite numerous. Adults emerge in the late spring and eat the foliage of a variety of plants.

Shining leaf chafers, subfamily Rutelinae, can be beautiful insects, but the subfamily has been given a bad name because it also contains the Japanese beetle and the oriental beetle, both alien species that have caused headaches for gardeners. Adult rutelines eat plant leaves or flower petals, while larvae develop either on grass roots or in rotting logs. If you live near a woodlot with rotting stumps, you may have noticed the large spotted pelidnota, a light tan beetle with six dark spots lining its wings.

The subfamily Dynastinae is not large, but the species in it are. In fact, the eastern Hercules beetle (*Dynastes tityus*) is the largest scarab in North America. The male sports two impressive horns: one on the top of his pronotum and the other on his head. Females are hornless. Children are invariably excited when they find these great beetles and often want to save them in a shoebox. (Unfortunately, the beetles are chockfull of oily fat that converts

The eastern Hercules beetle (*Dynastes tityus*) is the largest beetle in the United States. The male (top) is armed with powerful horns so he can fight other males for access to females. The female (bottom) has no reason to fight and thus has no horns to fight with.

The June beetle *Ochrosidea villosa* is common at night in our yard in early July.

Cotinus nitida, the green June beetle, is a rather large scarab that flies recklessly around your yard in early July as it searches for food and places to lay its eggs. Its large, heavy body can be difficult to steer while in flight and occasionally it will crash into you in a somewhat scary collision. Fear not. It is completely harmless.

the turquoise-green exoskeleton of the dead insect into a smelly and greasy black mess in a few days. A short bath in acetone can solve both problems; the color is fixed by the chemical and the fats are extracted from the body.) Dynastines complete their larval development in large, well-rotted logs.

The beautiful green June beetles (*Cotinus nitida*) belong to the subfamily Cetoniinae. Like other June beetles, the larvae of these rather large beetles develop on grass roots, and the green adults eat foliage. They are often seen flying back and forth over lawns in midsummer as they look for likely places to lay eggs. Gardeners occasionally dig the larvae up while working the soil. If you have encountered a *Cotinus* larva, you may have noticed as it crawled away that it was crawling on its back, with its feet in the air. A curious sight indeed!

Click beetles

Order Coleoptera, family Elateridae
Number of species North of Mexico: 965
Life stages: egg, larva, pupa, adult

Click beetles are common insects that all look pretty much the same, except for some variation in size and color. They are elongate, nearly linear beetles, and most species are brown and just under an inch in length, although our largest species are almost two inches long. Their most notable feature is their mode of defense. If you flip one on its back and look closely between the first pair of legs, you will notice a thumblike projection of the exoskel-

Alaus oculatus, our largest species of click beetle, sports two faux eyes on its prothorax. This display apparently convinces birds that this creature is more than they bargained for.

Although *Aeolus mellilus* is an attractive click beetle (left), *Limoneus propexus* better represents what most click beetles look like (above).

(below)The margined blister beetle (*Epicauta pestifera*) can be common on solanaceous plants like horsenettle (*Solanum carolinense*).

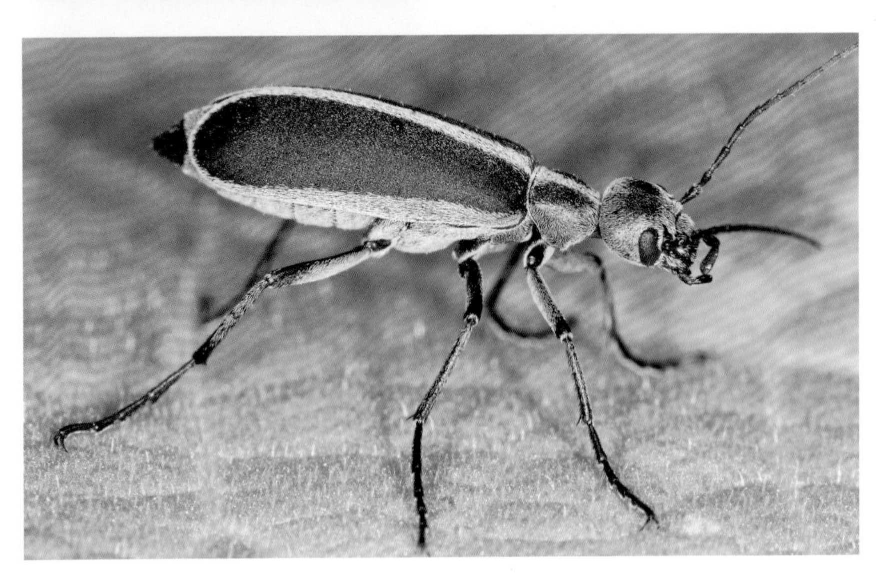

eton extending into a notch between the second pair of legs. When threatened, click beetles snap this projection with great energy into the notch. The force of this action is powerful enough to produce an audible "click" and launch the beetle several inches into the air. If the beetle lands on its back, it will "click" again, and it keeps clicking until it lands on its feet and scurries away.

Most click beetles develop in the ground, eating plant roots; but some in their larval stage, like the eyed click beetle (*Alaus oculatus*), are predators of other insects in rotting logs. Click beetle larvae are called wireworms and can be important food items for birds that forage in loose soil. Adults of all species are phytophagous.

Blister beetles

Order Coleoptera, family Meloidae
Number of species north of Mexico: 400 plus
Life stages: egg, larva, pupa, adult

Unlike most beetles, blister beetles are not protected by a tough exoskeleton but rather have a soft, leathery feel. Blister beetles are interesting creatures for several reasons. When threatened, they ooze blood from their joints in a process called reflex bleeding. If you handle a troubled beetle, some of this blood is likely to get on your skin. It's best to avoid this, because the blood contains cantharidin, a caustic compound that can raise blisters. Cantharidin is also a vasodilator that functions just like Viagra. An overdose can kill, but in old-time Europe, this side effect did not deter men eager to revisit their youth or, they thought, arouse their lady friends. A blister beetle species called the Spanish fly was dried, ground to a powder, and sold as an aphrodisiac for centuries. Like all other animal products sold as aphrodisiacs, it doesn't work (Rasor 1999).

Blister beetles eat foliage as adults, and several species have an affinity for solanaceous species like tomatoes and potatoes. As larvae, blister beetles from the genus *Epicauta* eat grasshopper eggs, and blister beetle larvae of other genera consume bee larvae. In my backyard, the black blister beetle (*Epicauta pennsylvanica*) is extremely common at the end of the summer on goldenrod flowers, where it eats the pollen.

Soldier beetles

Order Coleoptera, family Cantharidae
Number of species north of Mexico: 473
Life stages: egg, larva, pupa, adult

Soldier beetles got their name from the red and black coloration of several European species. They reminded somebody of British redcoats. Our North American species are typically tan, with dark brown patches on the tips of the wings. Like blister beetles, the exoskeleton of soldier beetles is soft and leathery. One might think that a beetle in the family Cantharidae would, like blister beetles, have cantharidin in its blood, but one would be wrong. Their blood is as sweet as yours and mine.

Soldier beetles are predators as larvae. I discuss them in this section on herbivores because the adults love pollen and spend much of their adult lives crawling over flowers in search of it. If you grow composites or asters, you have undoubtedly seen soldier beetles on your flowers. This is not a bad thing, for they eat aphids at every opportunity. Soldier beetles are also very common on goldenrod blooms at summer's end. These beetles spend much of their time mating. A female going about her business of eating pollen almost always has a male on her back and often will have two or three jockeying for the key position. At our place, soldier beetles are numerous early in July and then again in September.

Soldier beetles like *Chauliognathus pennsylvanicus* are very common on the flowers of composites in late summer.

Weevils and bark beetles

Order Coleoptera, family Curculionidae
Number of species north of Mexico: 3490 plus
Life stages: egg, larva, pupa, adult

Beetles are the most diverse animals on the planet, but weevils easily win the diversity contest among fellow beetles. Worldwide there are well over 40,000 species of weevils, and every one of them is a herbivore. The larval forms of a few species, like the alfalfa weevil, eat foliage very much like any caterpillar; but most weevils develop within plant tissue, be it stems, seeds, roots, rotting logs, or—in the case of bark beetles—in the cambium or heartwood of trees.

Evolutionary biologists are always trying to understand why things have evolved as they have, and one of the questions they have pondered for years is why weevils have diversified more than any other group. We have already discussed why being herbivorous increases species numbers. If each weevil species specializes on a particular plant group, at least as many weevil species as plant groups can evolve. But several other less diverse beetle families are also plant-eaters, so that explanation of weevil diversity is not sufficient alone. It is difficult to know for sure, but the favored hypothesis gives credit to the morphology of the weevil head. In most species the anterior (front) part of the head is elongated into what looks like a snout or nose. In some weevils the "snout" is broad and short; in others, it is ridiculously long, thin, and curved. Actually, it is not a snout at all, but a projection of the entire head. The weevil's mouth and all of its mouthparts are at the end of this projection, and that may be the key to its success. Beetles do not possess a sclerotized, or hardened, ovipositor (the structure through which eggs are laid), so they cannot insert eggs into protected areas very easily. Weevils have overcome this shortcoming in beetle design by using their snoutlike heads as ovipositors. No, they do not lay eggs through their mouths; but females do chew a tiny hole where they wish to lay an egg, deep within an acorn, for example, and then turn around and lay the egg in the hole. Beetles with very long snouts, like the pecan weevil, can bury their eggs at the center of a pecan, well out of harm's way. The ability to hide eggs not only enables weevils to protect their eggs from enemies; it also gives them the option of exploiting plant tissues that would otherwise be inaccessible to larvae. The egg-hiding ability gained

A weevil from Peru has greatly enlarged front legs for gripping branches while mating (left). Another species looks much like a bird dropping (right).

The weevil known as the cockleburr billbug (*Rhodobaenus tredecimpunctatus*) uses its elongated head (left) to chew a deep hole into a sunflower stem (right). It will then lay eggs in the hole, giving its larvae a food resource they would not be able to exploit on their own.

from the "snout" therefore partitions each plant lineage into niches available to weevils but not other beetles, setting up an advantage that may very well account for the fantastic diversity we find in the Curculionidae.

Moths and butterflies

Order Lepidoptera, 72 families
Number of species north of Mexico: nearly 12,000
Life stages: egg, larva, pupa, adult

When it comes to supplying food for other animals, no group of insects surpasses the moths and butterflies. All species in this large order are herbivores (with the minor exception of a few predaceous geometrids and pyralids, and a single lycaenid). The fat, juicy larvae are favorites of birds, rodents, and predaceous insects, and thus lepidopterans form a critical base of the food web that keeps ecosystems healthy. It follows that if we were forced to care for only one group of insects in our restored suburban ecosystem, we would do well to choose the Lepidoptera. Fortunately, moths and butterflies are not innately repulsive to most people, as are so many other insects. The tiny scales that are fixed to their wings create an aesthetic that humans appreciate, without training, from a very young age. Those of us who have watched adults or larvae closely are typically fascinated by the experience. I predict that few people will object to increasing the numbers and diversity of Lepidoptera species in our living spaces.

People often ask what differentiates moths from butterflies. Several generalities can be made, but most are fraught with exceptions. Most moths are nocturnal, while all butterflies are diurnal (active by day). Moths are typically fatter and the scales on their wings rub off on your fingers quite easily. Butterflies, in contrast, are more streamlined in the body and their wing scales are affixed more securely. Moth antennae are feathery (pectinate) or are narrow filaments without a terminal knob. Butterfly antennae always have a knob at the end that is recurved back like a small hook. Finally, moths usually hold their wings flat over their abdomen when resting, while butterflies hold the wings upright and pressed tightly together. Skippers seem to be a compromise between moths and butterflies. Their most distinguishing trait is that they hold their two wings at different angles when at rest; the hind wing is spread wider than the forewing.

There are 72 families of Lepidoptera in North America, most of which contain what taxonomists call microlepidoptera. Microleps are typically small, nondescript moths that fly only at night. They are the little white moths that as kids we kicked up out of the grass as we played hide and seek at dusk. They are the moths of various shades of brown that gather at our porch

Desmia funeralis, the grape leaffolder, is among the more beautiful microleps in the eastern United States.

The luna moth (*Actias luna*) prefers sweet gum in eastern forests.

The Virginia creeper sphinx (*Darapsa myron*) typifies the jet-fighter body plan of most sphinx moths.

lights or fly around our lanterns when we go camping. They are the insects that provide nutritious food for bats, nighthawks, and whippoorwills. Lepidoptera can be further divided into larger moths, skippers, and butterflies. These are categories of convenience rather than taxonomy.

Some of our more spectacular moths that cannot be considered microleps belong to the Saturniidae, the giant silk moth family. The beautiful cecropia, polyphemus, promethea, io, and luna moths belong to this group and are characterized by the elaborate cocoons they spin just before pupating. Species such as the royal walnut moth and the imperial moth also belong to this family of Lepidoptera, but these pupate in chambers underground and do not spin cocoons at all. Moths in the family Sphingidae are called sphinx moths as larvae, although adult forms are often known as hawk moths or hummingbird moths. These species have incredibly long, tubelike mouthparts that function as straws through which the adults suck nectar. The "straw" of a typical sphinx moth can be three times the length of its body and allows these moths to suck nectar from flowers with very deep corollas. Flowers that attract hummingbirds usually attract sphinx moths as well.

Like most other sphinx moth larvae, *Hemaris thysbe*, the hummingburd clearwing, has a "horn" on its rear end (left). Sphinx horns are harmless fleshy extensions that cannot harm you or slice open your tomatoes! Among the few species that do not have a horn is *Eumorpha pandorus*, the Pandora sphinx (right).

Species in the family Hesperiidae are called skippers because of their flight habits. Their short, quick flights among open vegetation make them look as if they are skipping across the field. Most species are brown and do not attract much attention, but they can be numerous in early successional habitats and are therefore important contributors to the stock of "bird food" during the summer months.

All of our butterfly species belong to one of four families: Lycaenidae, Pieridae, Papilionidae, or Nymphalidae. The lycaenids are all small butterflies and include the blues, the harvesters, the hairstreaks, the coppers, and the metalmarks. It is easy to encourage many species of lycaenids in suburbia because their hosts are common landscape plants. For example, the spring azure (*Celastrina ladon*) eats flower parts of dogwood, snakeroot, viburnum, and blueberry flowers. The harvester (*Feniseca tarquinius*) is exceptional in that it is the only North American butterfly that eats other insects rather than plants. Harvester larvae are specialists on woolly aphids that attack ash, beech, alder, hawthorn, and witch hazel. Having harvesters in your garden is definitely a good thing. The banded hairstreak (*Satyrium calanus*) devel-

Skippers have traits of both moths and butterflies. They are relatively heavy bodied like moths, fly during the day like butterflies, and have antennae unique to skippers. The larval form of *Anatrytone logan*, the Delaware skipper, develops on grasses.

ops on oaks, while the gray hairstreak (*Strymon melinus*) does best on the blossoms of legumes; massings of blue false indigo (*Baptisia australis*) or false indigo (*Amorpha fruticosa*) should encourage this common species, as well as species of sulphur butterflies (family Pieridae, *Colias* species). The American copper (*Lycaena phlaeas*) is a specialist on various species of introduced dock, especially sheep sorrel (*Rumex acetosella*). You may wonder how a butterfly from eastern North America could be a specialist on a group of alien plants. It can't! It turns out that the American copper is misnamed. It is actually a butterfly that was introduced from Europe accidentally along with its host plants during colonial times (Opler and Malikul 1992).

Swallowtails (family Papilionidae) are large butterflies that have a tail-like extension on each hind wing. Most species are specialists, which makes it easy to target the plants they need for development. The tiger swallowtail (*Papilio glaucus*) is our largest local species and is the most catholic in its host preference. Black cherry and tulip tree are its favorites, but it will take willow, ash, basswood, and birch as well. Female tiger swallowtails sometimes trade in their yellow and black coloration for an all-black form. This is

The butterfly family Lycaenidae includes smaller species like *Strymon melinus*, the gray hairstreak (A); *Lycaena phlaeas*, the American copper (B); *Everes comyntas*, the eastern tailed blue (C); and *Celastrina ladon*, the spring azure (D).

most common where the tiger swallowtail overlaps with populations of the pipevine swallowtail (*Battus philenor*). The pipevine swallowtail acquires a bitter taste from its eponymous host plant. Local birds learn that black butterflies taste bad, and so along with the pipevine swallowtail, they mistakenly avoid the dark form of the tiger swallowtail. Other true specialists in the Papilionidae include the zebra swallowtail (*Eurytides marcellus*), which eats only pawpaw; the spicebush swallowtail (*Papilio troilus*), which develops on spicebush and sassafras; and the black swallowtail (*Papilio polyxenes*), which requires members of the carrot family to complete development.

Swallowtails and their beautiful larvae are a welcome addition to any garden: *Papilio glaucus*, the tiger swallowtail (adult, A; larva,B); *P. polyxenes*, the black swallowtail (adult, C; larva, D); *P. troilus*, the spicebush swallowtail (adult,E; larva, F).

Danaus plexippus, the stately monarch, is one of the easiest butterflies to recruit to your butterfly garden. All you need is a milkweed patch.

The largest family of butterflies is Nymphalidae, which includes some 50 species in North America. The family comprises some of our most familiar butterflies, and many are easy to attract to suburban gardens. The monarch (*Danaus plexippus*), for example, can be encouraged and its numbers multiplied by planting any of several species of milkweeds. I have recently discovered four species of milkweeds on our property (green, common, swamp, and purple) and have added two more, butterfly weed and whorled milkweed, to produce a continuous bloom of milkweeds from June through September. Monarchs are a common sight at our house from July on, and their numbers increase as they lay more and more eggs on my milkweeds. I feel good about this, because as we've seen, the monarch is increasingly threatened by loss of overwintering habitat in the mountains of central Mexico. The more we can build monarch populations while they are in North America, the more likely the species will survive large losses during the winter months in Mexico.

As with other groups of insect herbivores, many nymphalids are quite specific about what they will accept as larval food, while others accept sev-

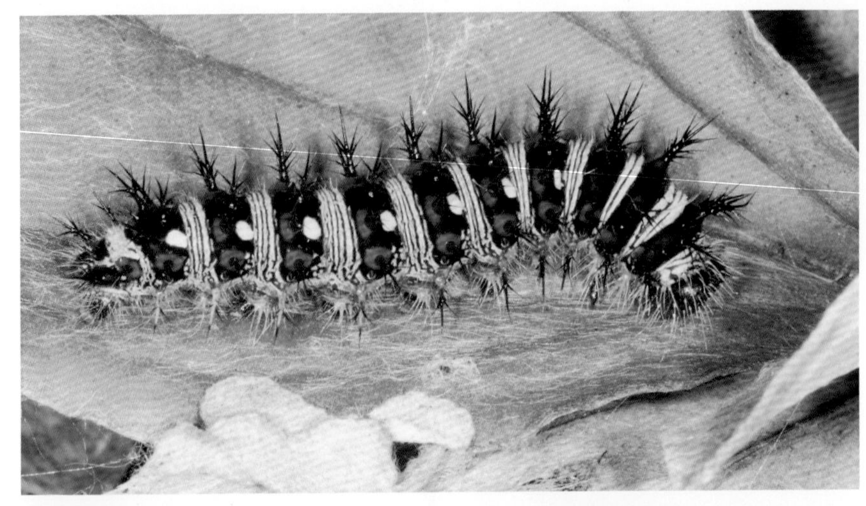
A

B

C

Satryodes erydice, the eyed brown, is one of the common wood nymphs seen flitting among trees in early summer.

The larvae of *Vanessa virginiensis*, the American lady (A), prefer as a host plant *Antennaria neglecta*, pussytoes (B). The adult American lady (C) looks very much like *Vanessa cardui*, the painted lady.

A

B

C

D

E

eral different plants. For example, the northern pearly eye (*Enodia anthedon*) will only eat grasses. The American lady (*Vanessa virginiensis*) does best on pussytoes (*Antennaria* species), while its cousin the painted lady (*Vanessa cardui*) chooses nettles (*Urtica* species), thistles (*Circium* species), asters, and hollyhocks. Several species of fritillaries, including the endangered regal fritillary (*Speyeria idalia*), eat only violets. It is easy to have healthy populations of the pearl crescent (*Phyciodes tharos*) by planting native asters. Both the question mark (*Polygonia interrogationis*) and the comma (*Polygonia comma*) can be supported by elm trees. Enthusiasts who want to see how many species of nymphalids they can sustain in their yard can easily expand this list to include red admirals, buckeyes, viceroys, hackberry butterflies, red-spotted purples, mourning cloaks, and various wood nymphs by planting the appropriate host plants.

Among the nymphalid butterflies that are easy to produce in your garden are *Vanessa atalanta*, the red admiral (A); *Speyeria cybele*, the great spangled fritillary (B); *Phyciodes tharos*, the pearl crescent (C); *Polygonia comma*, the comma (D); and *Polygonia interrogationis*, the question mark (E).

Sawflies

Order Hymenoptera, families Tenthredinidae, Pergidae, Argidae, Diprionidae, Cephidae, Cimbicidae, Siricidae, Xiphidriidae, Pamphiliidae, Xyelidae
Number of species north of Mexico: 1071
Life stages: egg, larva, pupa, adult

The sawflies constitute a group of herbivores that get little press. They are not flies at all, and even though they are true hymenopterans—whose species include wasps, bees, and ants—they act more like lepidopterans. Sawflies got their name because the ovipositor on females is serrated and looks a little like a saw. I still remember finding my first sawfly adult. I was taking insect taxonomy in graduate school, but we had not yet covered the order Hymenoptera. My specimen did not look like anything I had seen before. It was not a fly because it had four, rather than two, wings, and its wings were clear with visible veining. It did not look like a wasp because its abdomen was broadly joined to the thorax; that is, it didn't have a narrow "waist." Had I seen its larval form, I would have been even more confused. Sawfly larvae look just like Lepidoptera caterpillars, the only big difference being the number of fleshy "prolegs" they have on their abdomen. They also act like caterpillars and munch on leaves, either in groups or individually. This alone differentiates sawflies from all other Hymenoptera species: they are the only "wasps" that develop on leaves. Finally, because they are soft and usually taste good, birds love to feed them to their nestlings.

There are several families of sawflies, but the largest and by far the most common in this country are the Tenthredinidae. Over 800 species of tenthredinids occur in North America, but surprisingly little is known about their host plant preferences. Some of the first insects we encounter in the spring are adult tenthredinids. About the time red maples are in bloom and you are turning over your garden after its winter rest, tenthredinids start flying around, looking for places to lay their eggs. If you wear a yellow shirt, they will land on you as well, mistaking you for a flower. This should cause you no concern, as they neither sting nor bite.

The species we often encounter on alternate-leaf and redtwig dogwood belong to the genus *Macremphytus*. The insects feed gregariously until they are nearly mature. Young instars cover their bodies with a white powdery substance, probably to discourage predators. In some years, these sawflies

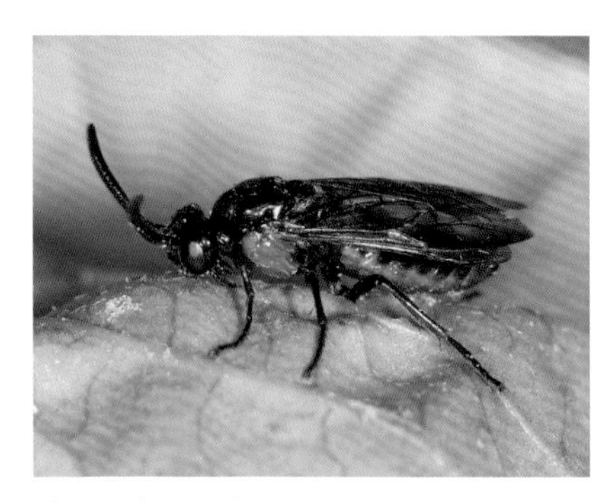

The sawfly, *Arge humoralis*, is one of the few insects that can develop on poison ivy. Sawfly larvae, like the Lepidoptera caterpillars for which they are easily mistaken, are an excellent source of food for wildlife.

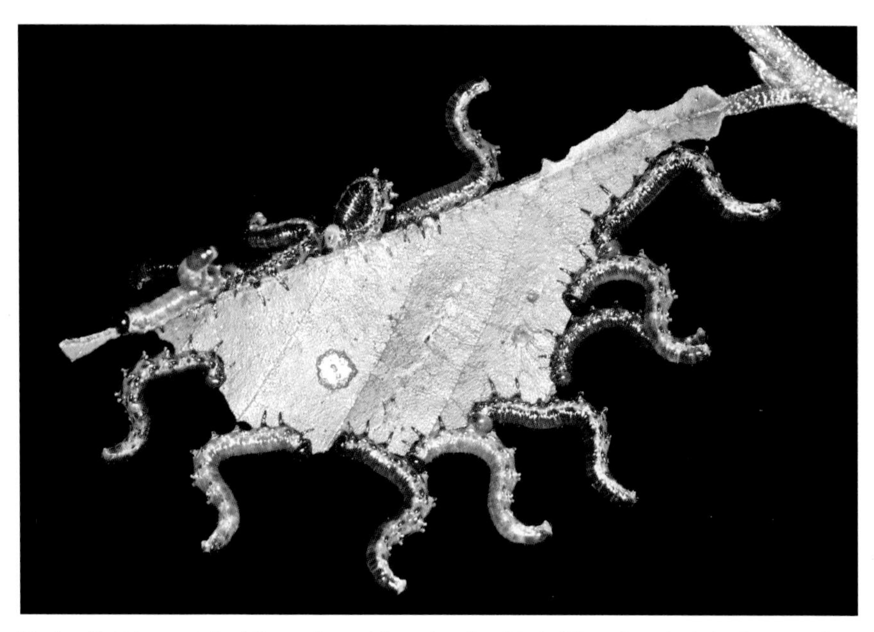

Tenthredinid larvae defend themselves while eating river birch. When disturbed, the larvae curl their abdomens over their bodies. This may make it more difficult for parasitoids to lay eggs in the sawflies.

seem to be everywhere in August. Some tenthredinids are leaf miners. The birch leaf miner (*Fenusa pusilla*) is an alien pest from Europe. The lack of its complement of adapted parasitoids has enabled it to spread throughout the Northeast, and it often turns entire birch trees brown. Three tenthredinid species form leaf or stem galls on various willow species. Most tenthred-

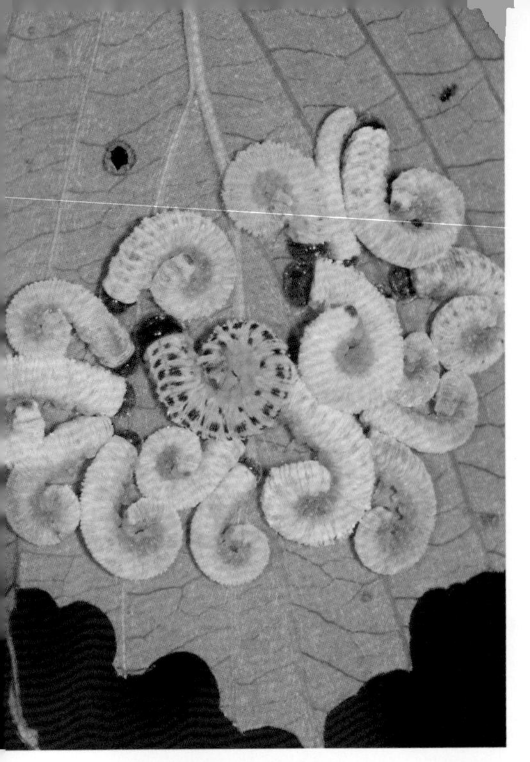

Macremphytus species can be numerous on alternate-leaf dogwood in August. They are gregarious when young (left) but move off on their own as last-instar larvae (right). Homeowners often do their best to rid their bushes of these tenthredinids, but at our house the birds do it for us. Early migrants feast on these sawflies as they move south.

The larva of the elm sawfly (*Cimbex americana*), one of our largest sawfly species.

inids, however, are external leaf-eaters on a wide variety of plant species. The more types of plants we have in the suburban landscape, the more bird food in the form of tenthredinid sawflies we will produce.

Although Tenthredinidae is the largest sawfly family, other families also make important contributions to sawfly numbers and diversity. *Cimbex americana*, a large sawfly, is in the family Cimbicidae and is found primarily on elm. Horntails (Siricidae) and wood wasps (Xyphidriidae) are sawflies that develop as larvae in wood. As one moves farther north, species in the Diprionidae and Pamphiliidae that specialize on various conifers become more common. Several of these have been introduced from Europe and are serious pests of northern forests. But diprionids taste good and are common enough to be important sources of protein for birds. When we have a cool, wet spring at our house, Lepidoptera larvae are small and hard to come by. Nevertheless, our resident family of bluebirds successfully fledges their young by relying on a steady diet of diprionid sawflies from our white pines.

ARTHROPOD PREDATORS

In my discussion of the importance of arthropods as food for other animals, I have focused on insect herbivores because they are the animals that pass most of the energy captured by plants to animals not able to eat plants directly. But I would be terribly remiss not to mention the degree to which arthropod predators of insect herbivores, particularly spiders and predatory insects, contribute to the diets of birds and other vertebrates. Most vertebrates that eat insects only discriminate among potential prey when there is a clear warning to do so. Usually, such warnings come in the form of an aposematic coloration: a color pattern that, as predators quickly learn, does not portend a tasty meal. If no such warning exists, however, a hungry bird snatches whatever flies by.

If the prey is itself a predator, it means that a plant's energy has passed through two trophic levels before reaching the bird, rather than just one. But that is immaterial as far as the bird is concerned. Nutritionally, predators are just as full of protein and high-energy fats as are herbivores, and the bird's hunger is just as satisfied. What I am saying here is that a large and diverse population of insect herbivores will generate a large and diverse population of insects and spiders that eat those herbivores. Vertebrates prey on the entire arthropod community, and the larger that community is, the

The larval form of the milkweed tiger moth (*Euchaetes egle*) is a great example of an insect herbivore protected by its orange and black aposematic coloration.

An inchworm (Geometridae) parasitized by tachinid flies and braconid wasps. The flies have laid their white eggs near the head of the inchworm, the wasps their cocoons in its abdomen.

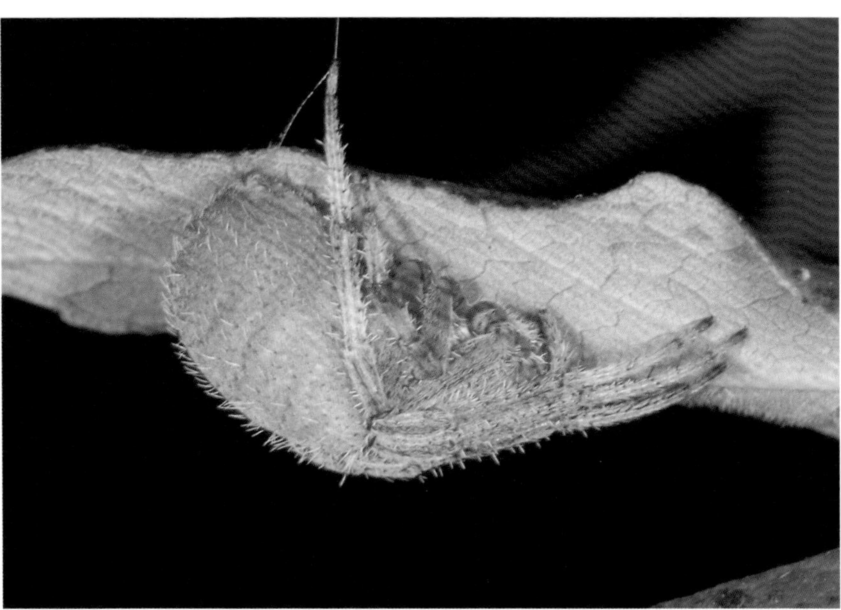

Spiders constitute up to 50 percent of the diet of some bird species that are rearing young.

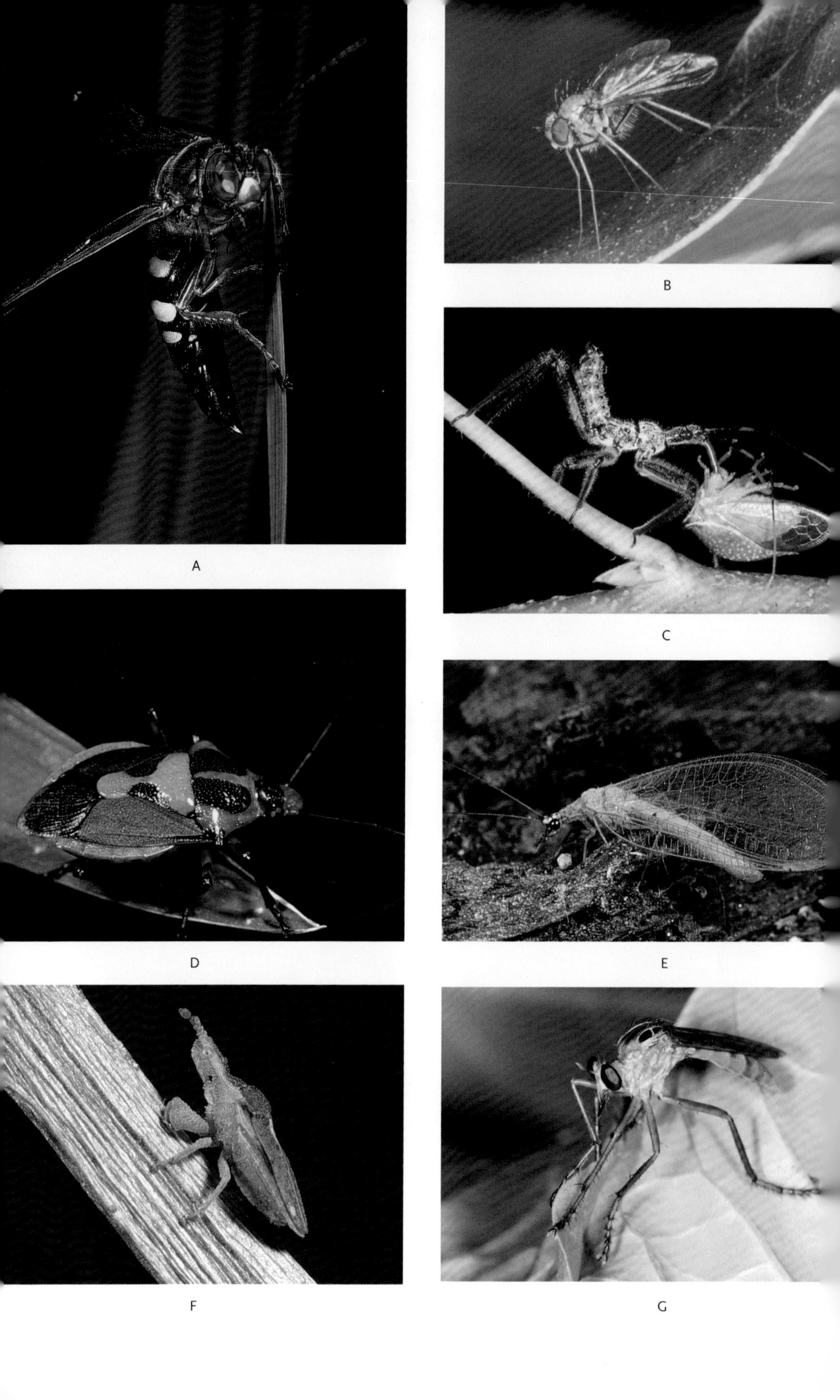
A

B

C

D

E

F

G

I

H

J

K

L

M

Among the many thousands of species of insect predators that should be part of a restored suburban ecosystem are (A) wasps (Specidae and Vespidae), (B) long-legged flies (Dolichopodidae), (C) assassin bugs (Reduviidae), (D) predatory stinkbugs (Pentatomidae), (E) lacewings (Chrysopidae), (F) ambush bugs (Phymatinae), (G, H) robber flies (Asilidae), (I) thick-headed flies (Conopidae), (J) praying mantids (Mantidae), (K) ladybird beetles (Coccinellidae), and (L) net-winged beetles (Lycidae). Also figuring in this account are countless species of parasitoids like (M) tachinid flies.

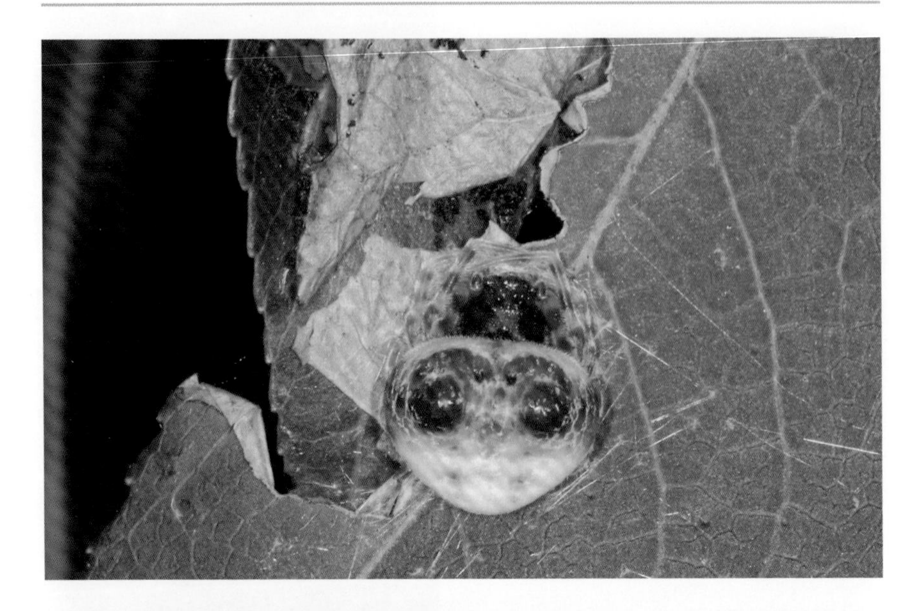

By day, bolas spiders in the genus *Mastophora* hide upon leaf splotches that conceal their bodies (above). By night, they call in moths by mimicking their sex pheromones (below).

more vertebrates there will be. So not only do the dragonflies, robber flies, and damsel bugs, the assassin bugs, ground beetles, and tiger beetles, the aphid lions, ladybird beetles, and minute pirate bugs, and the thick-headed flies, big-headed flies, and small-headed flies all help keep insect herbivore populations in bounds, they also help provide the stuff of life that enables vertebrate insectivores to exist.

And then there are the spiders. I didn't realize how much birds rely on spider populations to feed their young until I helped a colleague identify the prey items that house wrens bring to their nests. He had positioned a camera next to each nest box and took a picture of the wren each time it entered the house. My job was to look at these photos (thousands of them) and identify the arthropods in the wren's beak. The lesson I took home from that academic adventure was that spiders made up at least half of the prey items that wrens feed their young. Ornithologists tell me that this is the rule for many forest birds, such as the wood thrush, that forage much of the time on the ground. Although I understand the aversion many people feel toward spiders—a friend of mine used to throw them at me when I was five, and they have given me the creepy crawlies ever since—there is little doubt that spiders will play an important role in our restored suburban ecosystems, and this is as good a time as any to stop paying pesticide companies to eliminate them from our gardens.

I discovered one of the most fascinating spider species in our area doing its best to blend in with a leaf spot on our sweet gum tree. It was a bolas spider (*Mastophora* species), the first one I had ever seen. What makes these spiders interesting is the way they hunt (Yeargan 1988). By day, they rest over a leaf splotch and blend in very well. By night, instead of spinning a web, the bolas spider hangs motionless beneath its leaf, holding from one of its front legs a strand of silk with a glob of glue at its tip. When an insect flies by, the spider swings the bolas, just as human hunters used to do, and snags the unlucky victim on its sticky end. You can guess what happens next. What I find particularly neat is that bolas spiders do not leave encounters with prey up to chance. The probability of something flying right to the spider must be pretty low. So, over evolutionary time, the spider has specialized in capturing particular species of moth. To attract this moth, the bolas spider releases a pheromone that mimics the moth's sex pheromone. Male moths, thinking they are on their way to an evening tryst, find themselves at the sticky end of a bolas instead. Even more amazing is that a single species of bolas spider can mimic the pheromone of more than one moth species. The spider releases the appropriate pheromone at the time of night that particular species of moth takes flight. Now who could step on such a creature?

CHAPTER FOURTEEN

Answers to Tough Questions

When I talk to groups of gardeners about the use of native plants to sustain our nation's biodiversity, several questions pop up repeatedly. This chapter addresses some of the most common of those questions.

Q: Why can't we let nature take its course and just leave the aliens alone?

Unfortunately, nature is not as all-powerful as we might wish. "Nature" can be defined as the plants and animals in a given area and all of the natural phenomena that made them as they are. Through eons of evolution by natural selection, living things adapted to their physical environment and the organisms around them in ways that enabled them to survive and pass their genes on to future generations. These natural processes worked well within the ancestral setting that created them, but we humans have changed that setting drastically—and almost instantly, when viewed on an evolutionary time scale. The plants and animals in today's world have had no time at all to adapt to these sudden changes and so are still operating under the rules that

A "natural area" in Cape May, New Jersey, is not able to proceed through secondary succession to mature hardwood trees because the alien porcelain berry at the site smothers any native seedling that germinates. The vines cover young sweet gum trees that will not fulfill their role in secondary succession unless the porcelain berry is removed.

worked before humans took over landscape management. The end result is that without direct intervention by the humans who have placed them at risk, most organisms will not survive under our rules.

We have allowed alien plants to replace natives all over the country. Our native animals and plants cannot adapt to this gross and completely unnatural manipulation of their environment in time to negate the consequences. Their only hope for a sustainable future is for us to intervene to right the wrongs that we have perpetrated. In order to let nature take its course, we must first *recreate* nature.

If birds eat the berries of alien plants like autumn olive, multiflora rose, and oriental bittersweet, why shouldn't I plant those species?

It's true that many aliens, particularly some of our most invasive aliens, produce large crops of edible berries. In fact, one of the reasons these species are invasive is that birds do eat their berries and then poop out the seeds miles from the parent plant. The problem is that berries only support birds after the birds reproduce each season; they provide nothing for the nestlings. Many migrants rely in part on berries for energy just before and during their flights south, and nonmigrants eat berries and their seeds throughout the long winter months. But during reproduction, most bird species depend on insect protein and fat for the nutrients required to make eggs, for their own energy needs while feeding young, and—most important—for the young themselves. If having aliens in your yard means there are fewer insects available for birds during vital periods of reproduction, it doesn't matter how many berries are present afterward; there won't be a new crop of birds to eat them! And wherever did we get the idea that the berries from alien plants are better for our birds than the berries from native plants? If we want berries in our yards, why not plant arrowwood (*Viburnum dentatum*), elderberry (*Sambucus canadensis*), alternate-leaf dogwood (*Cornus alternifolia*), Virginia creeper (*Parthenocissus quinquefolia*), spicebush (*Lindera benzoin*), red or black chokeberry (*Aronia* species), winterberry (*Ilex verticillata*), various native hawthorns (*Crataegus* species), or red cedar (*Juniperus virginiana*)? What do we think the birds were eating before we took it upon ourselves to plant what they "need"? In the past, our eagerness to help got a little ahead of the knowledge we needed to succeed. We know better now, so let's avoid these mistakes in the future.

Because they produce copious berries, aliens like the autumn olive (A) have been planted to help birds. We can give birds not only lots of berries but the insects they need to feed their young by planting species like alternate-leaf dogwood (B), elderberry (C), and arrowwood (D).

A kingbird prepares to enjoy a meal of alternate-leaf dogwood berries.

Q: Isn't habitat destruction a more pressing problem than alien plants in the landscape?

Habitat destruction as a result of anthropogenic changes is a huge problem everywhere for life on earth. That is precisely why we can no longer rely on natural areas alone to provide food and shelter for biodiversity. Instead, we must restore native plants to the areas that we have taken for our own use so that other species can live along with us in these spaces. We can start by restoring native plants to our gardens. This is a manageable task for both suburban and city dwellers, with tangible results in a few short seasons as individual gardens begin to attract the birds and the insects that will sustain them. Just imagine the restored landscape that could result from everyone's cumulative efforts!

Q: Today there are more and more showy cultivars of native species on the market. Is it all right to use them, or will insects treat them like aliens?

I predict that in most cases, cultivars of native plants should be fine. The studies to prove this have not been done, but the chances are slim that a genetic change in the flower color or the fall leaf coloration of a native species would substantially change the palatability of that genotype for native insect specialists. I applaud any enhancements of natives that make them more attractive for homeowners and would encourage nurseries to do whatever it takes to bring as many native plant species into mass production as possible. One note of caution: I have heard that selections that increase berry size over the natural state may end up creating berries too large for some birds to eat. Some of the cultivars of winterberry (*Ilex verticillata*) may be a good example. But I reiterate that this is hearsay and has not yet been studied with controlled experiments.

Q: What's wrong with leaving vines on trees? I've heard they are good for the birds.

Some native vines like poison ivy, Virginia creeper, and fox grape do produce fruits that birds eat in the fall and winter (as do oriental bittersweet and Japanese honeysuckle, the scourges of the East). So in one sense, our native vines function as good neighbors in their ecosystems, providing both insects

The weight of vines on edge trees can be enormous and almost always results in the early death of the tree (left). Vines also girdle young trees. Even if the vine is removed (right), the tree is permanently weakened.

and fruit for birds and other wildlife. Today, however, there is a serious problem with woody vines, regardless of their geographical origin, because of the way we have fragmented the eastern deciduous forest. In the old days, woody vines were restricted to edge habitats where there was plenty of sun. When a tree fell down or a forest fire opened a gap in the middle of the woods, viney species would use the trees on the edge of the opening for support as they climbed up into the canopy, where sunlight was always available. When the forest gap closed, the seeds from the vines no longer germinated, or if they did, the seedlings were unable to thrive in the shade of the closed canopy. So vines were ephemeral in their impact on the trees of a mature forest.

Today, humans have inverted the spatial design of forests. Whereas most forests used to be continuous woods with closed canopies, now most of the land is cleared of trees and our forest patches are so small that they are nearly all edge habitats ideal for vine growth. Trees growing in these edge habitats are exposed to vine growth their entire lives rather than just for a short period. An overload of vines can damage trees in three ways. All of the woody vine species in the East grow faster than our trees, so trees end up being smothered (photosynthesis is blocked), girdled, or eventually pulled down by the sheer weight of the vines. Oriental bittersweet is particularly good at killing our native trees. It easily grows 10 feet or more each year and its woody stems wind their way tightly up the trunks and along the branches of our native trees, girdling them as the tree grows. If the girdling doesn't kill the tree, the weight of the vine does. It is now difficult to drive in the Piedmont of Maryland, Pennsylvania, or Delaware without seeing large trees such as sycamores, black cherries, oaks, and tulip trees snapped off at midheight by the weight of oriental bittersweet. Trees strong enough to support large vines during the summer are vulnerable during the winter when snow and ice accumulate on their vine-laden branches. Japanese honeysuckle and our native fox grape are also problems for trees these days, for the same reason. Our destruction of continuous forest habitats has thrown the ecosystem out of balance. The solution, of course, is for us to manage vine growth. To give the trees a chance, we must rip out every oriental bittersweet we see—is my passion showing?—and reduce the number of grapevines on our properties.

Q: If insects don't like alien plants, why do I see so many bees and butterflies at my butterfly bush?

When I claim that insects don't like alien plants, I am referring to insect herbivores: the leaf-eaters that necessarily encounter leaf chemical defenses while going about their business. Insects that drink nectar or eat pollen are not generally exposed to secondary metabolic defenses. Alien flowers are attractive to nectar lovers for a couple of reasons. First, nectar is almost entirely sugar and water. There are subtle chemical differences in nectar among plant species, but by and large, nectar from alien plants is the same as nectar from native plants. What are more important to bees than nectar

Natives like ground-nesting andrenid bees prefer to forage on native flowers.

constituents are flower shape and the amount of nectar stored in each flower. Each species of native bee evolved to forage in flowers with particular morphologies. Bumblebees, for example, have long "tongues" that can reach nectar pools at the base of flowers with long corollas, while sweat bees have relatively short tongues. This is why sweat bees spend most of their time foraging for nectar on flowers like composites, where the nectar is within easy reach, while bumblebees go for nectar in flowers with long necks where the nectar is protected from the short-tongued bees. If we plant an alien with long-necked flowers, our native bumblebees are already adapted to forage in those flowers. Butterflies all have long tongues and can forage in any flower that houses enough nectar to make it worth their while.

That said, there is growing evidence that our native bees, the andrenids, halictids, colletids, anthophorids, and megachilids, prefer native flowers to alien flowers. In her master's thesis at the University of Delaware, Nicole Cerqueira compared the number of visits of native bees to native and alien flowers in the same plant genus. She found, for example, that leaf-cutter bees (Megachilidae) visited *Lysimachia ciliata,* a native plant, at a rate of 3 bees every 10 minutes. During the same period, these bees never visited *L. nummularia,* an alien placed within inches of the native. She obtained the same results when she compared andrenid visits to native and alien *Geranium* species. In all, Nicole found that native bees showed a statistical preference for native flowers in 31 comparisons and never preferred alien flowers (Cerqueira 2005).

Another reason you may see lots of bees at your alien flowers is that honeybees, the most common bees in most areas of the country, are themselves alien creatures. Before Europeans brought them here there were no honeybees in North America. Since most of our alien plants have hailed from Europe or Asia where honeybees are native, the interaction between them is a natural one.

Q: My house sits on an eighth of an acre. Is that enough land to make a difference if I use natives instead of aliens?

Your small plot is connected to other plots, which are connected to others and others and others. Collectively they are North America. Changing the plant base of all of suburbia is quite an undertaking, but all *you* have to worry about is your eighth of an acre. Planting the back and side borders of your lot will provide more habitat than you might think, especially if you can get your neighbors to do the same. If your plantings are 15 feet wide, and your neighbor's border plantings are also 15 feet wide, together you have created a 30-foot swath of habitat for the length of your yard, as well as a privacy screen that can enhance the value of your property. The important thing to remember is that even if you seem like the only one in all of North America who uses more natives than aliens, wildlife will be better off for your efforts. The effects will be cumulative, and probably synergistic, as more and more people join you. And don't forget that plants are long-lived. The white oak you plant tomorrow could easily live 300 years, servicing innumerable insects, birds, squirrels, mice, raccoons, and deer every year of its life. Yes, you can make a difference on a small plot of land. You can even make a difference if you own no land. If you live in an apartment, you may be able to influence the landscaping habits of your landlord, or the company you work for, or the township supervisors who control your city parks, or your sibling who does own property. If we humans are capable of turning hundreds of millions of acres of rainforest into depleted grasslands, and extirpating millions of buffalo from the plains, and billions of passenger pigeons from the skies and cod from the North Atlantic, we are also capable of returning natives to our gardens.

You can generate considerable amounts of habitat by generously planting your property lines with native shrubs and trees.

Q: I have a terrible problem with deer eating everything I plant. Should I buy plants marketed as "deer proof"?

It certainly is tempting to plant only species that deer won't touch. The extraordinary deer populations that now roam suburbia through much of the Northeast have all but ended recruitment of young plants into our woodlots and yards by eating their leaves in the summer and their woody tissue in the winter. Although there are claims that deer prefer aliens to natives (Parker, Burkepile & Hay 2006), you couldn't prove it by the deer at my house. There are precious few natives that deer don't love. This means that most plants sold as being unpalatable to deer are unpalatable because they

White-tailed deer in many parts of the East are putting terrific pressure on suburban gardens. These deer are part of a herd of 11 that do their best to eat all of our young trees.

are really unpalatable aliens. The deer are a problem because we have created endless edge habitat—ideal for deer proliferation—while eliminating all deer predators. The only things slowing the population growth of deer at all in the suburbanized landscapes of the East are cars and starvation. Neither is a suitable or compassionate solution to deer overpopulation. But replacing palatable natives in the landscape with unpalatable aliens is not the answer either. Single-strand solar-powered electric fences are said to provide effective protection in some areas, although I have not tried this yet myself.

If alien plants are not as good for birds as native plants, why do I always see birds flitting in and out of the aliens, and why do birds use aliens for nest sites?

Birds need three things to survive: protection from predators, nest sites, and food. Birds are not plant taxonomists. They are not innately prejudiced against alien plant species and do not avoid them on principle. Birds respond to plants based on their needs at the moment. If a sharp-shinned hawk flies by, the juncoes and chickadees in the area will dive into the nearest and densest vegetation for cover, regardless of where that vegetation evolved. In an emergency, any plant that provides the needed protection will do. Alien multiflora rose, buckthorn, and autumn olive are often now the only low bushes available, so those are the species in which birds hide.

Birds choose nest sites in the same way. A chipping sparrow, for example, prefers to nest in the dense shrubs of open fields. On my southeastern Pennsylvania stomping grounds, this criterion is often met by multiflora rose and autumn olive, which have replaced native blackberries and red cedar in many places. Chipping sparrows in such areas have no choice but to nest in multiflora rose, where they find the physical conditions needed for their nesting.

The mother of these baby catbirds chose to build her nest in a forsythia bush, not because the forsythia provides something native plants do not, but because it was the only bush in the yard.

Birds, however, will not be in our future if we provide them only with shelter and nesting sites. Their most important requirement is food, and as we have seen, most alien plant species do not produce the insects birds need to feed their young—to make more birds. Meeting two out of three of a bird's most pressing needs is just not enough. I cannot imagine why any bird lover would insist on using a plant that provides ⅔ of the services required by birds when there are dozens of native species that provide all of what birds need—cover, nest sites, and food.

Q: If an alien plant has been in this country long enough, doesn't it become a native?

People often equate the ability to naturalize with "becoming native," but they are very different things. There is no doubt that many alien plants have the ability to invade and become established in our ecosystems; that is, become naturalized. There is more to being "native," however, than being able to live in natural ecosystems with no horticultural assistance from humans. To understand what it takes to be native, we first have to remember what it is to be alien. I have argued that a plant is alien to a particular region when it has not shared an evolutionary history with the plants and animals in that region. Yet if an alien plant could fit seamlessly into an evolutionarily novel ecosystem by performing all of the roles that the native plants it has displaced used to perform, wouldn't it be a "native" functionally? Yes, it would. So the question then becomes, "How long does it take an alien plant species to become the ecological equivalent of the plants it has displaced?" Surely there are alien species that photosynthesize at the same rate as our natives, and hold soil as well as our natives, but this book has focused on how well alien plants pass food to animals in other trophic levels. This is accomplished by being palatable to our native herbivores, in particular, our insects. Do alien plants become more palatable to native insects over time? Theory predicts that native insects should eventually adapt to the secondary metabolic compounds found in alien plant leaves. How long does this take? If native insects could adopt alien plants for their growth and reproduction in just a few years, then aliens truly could become "natives" and we would have little to worry about. Unfortunately, most evidence suggests that this adaptation happens too slowly to make a difference.

Examining the insect fauna on a plant in the country in which it evolved gives us a pretty good idea of how many insect species that particular plant is capable of supporting. If we compare that number with the number of insects supported by the same plant species in the United States, we will have a direct measure of how well our insects are using the nutrition within that plant's tissues, that is, how well our insects have taken advantage of the food niches available on that plant. In the table below, I revisit data on the hosting capacity of various exotic species now widespread in North America. Note that even hundreds of years after a plant has been introduced to a novel geographic area—in these cases, the United States—it supports only a tiny fraction of the insects that it supports in its homeland. The evolutionary interactions that knit each of these plants tightly into the fabric of their native ecosystems have only just begun to develop in this country. It may be 10 thousand years, or many hundreds of thousands of years, before these plants play the same roles in the food web in North America that they play whence they came.

HOSTING CAPACITY OF ALIEN PLANTS INTRODUCED TO NORTH AMERICA

Plant Species	Herbivores Supported in Homeland	Herbivores Supported in North America	Years Since Introduction to North America	Reference
Clematis vitalba	40 species	1 species	100	Macfarlane & van den Ende 1995
Eucalyptus stellulata	48 species	1 species	100	Morrow & La Marche 1978
Melaleuca quinquenervia	409 species	8 species	120	Costello et al. 1995
Opuntia ficus-indica	16 species	0 species	250	Annecke & Moran 1978
Phragmites australis	170 species	5 species	300+	Tewksbury et al. 2002

AFTERWORD

The Last Refuge

I have attempted to make several points in this book, but they all converge on a common theme: we humans have disrupted natural habitats in so many ways and in so many places that the future of our nation's biodiversity is dim unless we start to share the places in which we live—our cities and, to an even greater extent, our suburbs—with the plants and animals that evolved there. Because life is fueled by the energy captured from the sun by plants, it will be the plants that we use in our gardens that determine what nature will be like 10, 20, and 50 years from now. If we continue to landscape predominantly with alien plants that are toxic to insects—the most important herbivores in our suburban ecosystem in terms of passing energy from plants to other animals—we may witness extinction on a scale that exceeds what occurred when a meteor struck the Yucatan peninsula at the end of the Cretaceous period. If instead we use plants that evolved with our local animal communities as the foundation of our landscapes, we may be able to save much of our biodiversity from extinction. In essence, we will for the first time coexist with nature rather than compete with her.

For the past century we have created our gardens with one thing in mind: aesthetics. We have selected plants for landscaping based only on their beauty and their fit within our artistic designs. Yet if we designed our buildings the way we design our gardens, with only aesthetics in mind, they would fall down. Just as buildings need support structures—girders, I-beams, and headers—to hold the graceful arches and beautiful lines of fine architecture in place, our gardens need native plants to support a diverse and balanced food web essential to all sustainable ecosystems.

A tiger swallowtail enjoys a Joe-Pye weed planting.

To me the choice is clear. The costs of increasing the percentage and biomass of natives in our suburban landscapes are small, and the benefits are immense. Increasing the percentage of natives in suburbia is a grassroots solution to the extinction crisis. To succeed, we do not need to invoke governmental action; we do not need to purchase large tracts of pristine habitat that no longer exist; we do not need to limit ourselves to sending money to national and international conservation organizations and hoping it will be used productively. Our success is up to each one of us individually. We can each make a measurable difference almost immediately by planting a native nearby. As gardeners and stewards of our land, we have never been so empowered—and the ecological stakes have never been so high.

APPENDIX ONE

Native Plants with Wildlife Value and Desirable Landscaping Attributes by Region

NEW ENGLAND

Maine, New Hampshire, Vermont, Massachusetts, Rhode Island, Connecticut

Shade and specimen trees

Acer nigrum, black maple
Acer rubrum, red maple
Acer saccharum, sugar maple
Betula alleghaniensis, yellow birch
Betula lenta, sweet birch
Betula nigra, river birch
Betula papyrifera, paper birch
Betula populifolia, gray birch
Carya ovata, shagbark hickory
Diospyros virginiana, persimmon
Fagus grandifolia, American beech
Fraxinus americana, white ash
Fraxinus nigra, black ash
Juglans cinerea, butternut
Juglans nigra, black walnut
Liquidambar styraciflua, sweet gum
Liriodendron tulipifera, tulip tree
Nyssa sylvatica, black gum
Oxydendron arboreum, sourwood
Platanus occidentalis, American sycamore

Populus deltoides, eastern cottonwood
Prunus serotina, black cherry
Quercus alba, white oak
Quercus bicolor, swamp white oak
Quercus coccinea, scarlet oak
Quercus macrocarpa, bur oak
Quercus muehlenbergii, chinkapin oak
Quercus palustris, pin oak
Quercus prinus, chestnut oak
Quercus rubra, red oak
Robinia pseudoacacia, black locust
Sassafras albidum, sassafras
Tilia americana, basswood

Shrub and understory trees

Acer penslyvanicum, striped maple
Amelanchier arborea, downy serviceberry, shadblow serviceberry
Aronia arbutifolia, red chokecherry
Aronia melanocarpa, black chokeberry
Carpinus caroliniana, ironwood
Ceanothus americanus, New Jersey tea
Celtis occidentalis, common hackberry
Cephalanthus occidentalis, buttonbush
Cercis canadensis, redbud
Clethra alnifolia, sweet pepper bush
Cornus alternifolia, alternate-leaf dogwood
Cornus florida, flowering dogwood
Cornus sericea, redtwig dogwood
Corylus americana, American hazelnut
Crataegus crus-galli, cockspur hawthorn
Crataegus mollis, downy hawthorn
Crataegus punctata, thicket hawthorn
Diervilla lonicera, northern bush honeysuckle
Dirca palustris, leatherwood
Gaylussacia baccata, black huckleberry
Hamamelis virginiana, witch hazel
Ilex glabra, inkberry
Ilex opaca, American holly
Ilex verticillata, winterberry
Kalmia latifolia, mountain laurel
Lindera benzoin, spicebush
Magnolia virginiana, sweetbay magnolia
Myrica pensylvanica, northern bayberry
Nemopanthus mucronatus, American mountain holly
Ostrya virginiana, American hop hornbeam
Physocarpus opulifolius, common ninebark
Prunus americana, American plum
Prunus virginiana, chokecherry
Ptelea trifoliata, hoptree
Rhododendron canadense, rhodora
Rhododendron maximum, great laurel
Rhododendron periclymenoides, pinxter azalea
Rhododendron prinophyllum, roseshell azalea
Rhododendron viscosum, swamp azalea
Rhus aromatica, fragrant sumac
Rhus copallina, winged sumac

Rhus glabra, smooth sumac
Rhus typhina, staghorn sumac
Rosa carolina, Carolina rose
Rosa virginiana, Virginia rose
Rubus idaeus, red raspberry
Rubus odoratus, flowering raspberry
Rubus pubescens, dwarf blackberry
Salix petiolaris, meadow willow
Sambucus canadensis, elderberry
Shepherdia canadensis, russet buffaloberry
Sorbus americana, American mountain ash
Spiraea alba, white meadowsweet
Spiraea tomentosa, steeplebush
Staphylea trifoliata, bladdernut
Symphoricarpos albus, snowberry
Vaccinium angustifolium, lowbush blueberry
Vaccinium corymbosum, highbush blueberry
Viburnum acerifolium, mapleleaf viburnum
Viburnum dentatum, arrowwood
Viburnum lentago, nannyberry
Viburnum nudum, smooth witherod, possumhaw
Viburnum prunifolium, blackhaw
Viburnum trilobum, American cranberrybush
Zanthoxylum americanum, pricklyash

Conifers

Abies balsamea, balsam fir
Chamaecyparis thyoides, Atlantic white cedar
Juniperus virginiana, red cedar
Larix laricina, tamarack
Picea glauca, white spruce
Pinus banksiana, jack pine
Pinus resinosa, red pine
Pinus rigida, pitch pine
Pinus strobus, white pine
Taxus canadensis, Canada yew
Thuja occidentalis, arborvitae, northern white cedar

Vines

Celastrus scandens, American bittersweet
Lonicera dioica, limber honeysuckle
Menispermum canadense, moonseed
Parthenocissus quinquefolia, Virginia creeper
Parthenocissus vitacea, woodbine
Vitis aestivalis, summer grape
Vitis labrusca, fox grape
Vitis riparia, riverbank grape

Streamside plants

Alnus incana, gray alder
Alnus viridis, mountain alder
Betula pumila, bog birch
Cephalanthus occidentalis, buttonbush
Chamaedaphne calyculata, leatherleaf
Clethra alnifolia, sweet pepper bush
Cornus amomum, silky dogwood
Dirca palustris, leatherwood
Eubotrys racemosa, swamp doghobble
Ilex verticillata, winterberry

Salix amygdaloides, peachleaf willow
Salix candida, hoary willow
Salix discolor, pussy willow
Salix interior, sandbar willow
Salix nigra, black willow

Ground covers

Asarum canadense, wild ginger
Cornus canadensis, bunchberry
Epigaea repens, trailing arbutus
Gaultheria procumbens, teaberry, wintergreen
Gaylussacia brachycera, box huckleberry
Juniperus horizontalis, horizontal juniper
Mitchella repens, partridgeberry
Phlox subulata, moss pink
Podophyllum peltatum, mayapple
Sibbaldiopsis tridentata, shrubby fivefingers
Vaccinium vitis-idaea, mountain cranberry
Waldsteinia fragarioides, barren strawberry

Herbaceous perennials, dry sites

Anemone canadensis, Canada anemone
Aquilegia canadensis, wild columbine
Asclepias syriaca, common milkweed
Asclepias tuberosa, butterfly weed
Aster divaricatus, white wood aster
Aster ericoides, white heath aster
Aster novi-belgii, New York aster
Caulophyllum thalictroides, blue cohosh
Chimaphila maculata, spotted wintergreen
Cypripedium parviflorum, lesser yellow lady's slipper
Eupatorium rugosum, white snakeroot
Geranium maculatum, wild geranium
Hedyotis caerulea, bluets
Heliopsis helianthoides, oxeye
Hepatica acutiloba, sharp-lobed hepatica
Hydrastis canadensis, goldenseal
Lilium philadelphicum, wood lily
Lobelia siphilitica, great blue lobelia
Panax quinquefolius, ginseng
Rudbeckia hirta, black-eyed Susan
Rudbeckia laciniata, cutleaf coneflower
Sanguinaria canadensis, bloodroot
Sisyrinchium angustifolium, blue-eyed grass
Solidago caesia, blue-stem goldenrod
Soldago rugosa, rough-stem goldenrod
Tiarella cordifolia, eastern foamflower
Tradescantia ohiensis, spiderwort
Trillium erectum, purple trillium
Trillium grandiflorum, white trillium
Uvularia grandiflora, large-flowered bellwort
Viola adunca, hookspur violet
Viola pedata, birdsfoot violet

Herbaceous perennials, moist and wet sites

Actaea rubra, red baneberry
Anemone quinquefolia, wood anemone
Arisaema triphyllum, Jack-in-the-pulpit
Asclepias incarnata, swamp milkweed
Aster novae-angliae, New England aster
Calla palustris, water arum
Chamaelirium luteum, devil's bit
Chelone glabra, turtlehead
Cimicifuga racemosa, black cohosh
Clintonia borealis, blue bead lily
Coreopsis rosea, pink coreopsis
Dicentra canadensis, squirrel corn
Dicentra cucullaria, Dutchman's breeches
Erythronium americanum, yellow trout lily
Eupatorium maculatum, spotted Joe-Pye weed
Eupatorium perfoliatum, boneset
Gentiana clausa, closed gentian
Helenium autumnale, Helen's flower, sneezeweed
Hepatica americana, round-lobed hepatica
Hibiscus moscheutos, rose mallow
Iris prismatica, slender blue flag
Iris versicolor, blue flag iris
Lilium canadense, Canada lily
Lilium superbum, Turk's cap lily
Lobelia cardinalis, cardinal flower
Lupinus perennis, blue lupine
Monarda didyma, beebalm
Monarda fistulosa, wild bergamot
Penstemon digitalis, foxglove beardtongue
Penstemon hirsutus, hairy beardtongue
Phlox divaricata, wild blue phlox
Phlox maculata, wild sweet William
Physostegia virginiana, false dragonhead
Polygonatum commutatum, Solomon's seal
Sanguisorba canadensis, American burnet
Smilacina racemosa, false Solomon's seal
Teucrium canadense, wood sage
Tradescantia virginiana, Virginia spiderwort
Veronicastrum virginicum, Culver's root
Viola cucullata, marsh blue violet
Viola pubescens, yellow violet
Zizia aurea, golden Alexanders

Grasses, sedges, and rushes

Andropogon gerardii, big bluestem
Andropogon virginicus, broomsedge
Carex pensylvanica, Pennsylvania sedge
Carex platyphylla, broadleaf sedge
Deschampsia cespitosa, tufted hairgrass
Eragrostis spectabilis, purple lovegrass
Juncus effusus, common rush
Panicum virgatum, switchgrass
Scirpus cyperinus, bulrush, woolgrass
Schizachyrium scoparium, little bluestem

Spartina pectinata, prairie cordgrass

Sporobolus heterolepis, prairie dropseed

Tridens flavus, purpletop, redtop

Typha angustifolia, narrow-leaf cattail

Ferns

Adiantum pedatum, maidenhair fern

Asplenium platyneuron, ebony spleenwort

Asplenium scolopendrium, American hart's tongue fern

Athyrium filix-femina, lady fern

Botrychium virginianum, rattlesnake fern

Cryptogramma stelleri, slender rock brake

Cystopteris fragilis, brittle bladder fern, fragile fern

Dennstaedtia punctilobula, hayscented fern

Dryopteris cristata, crested wood fern

Dryopteris filix-mas, male fern

Dryopteris goldiana, Goldie's fern

Dryopteris intermedia, evergreen wood fern

Dryopteris marginalis, marginal shield fern

Gymnocarpium dryopteris, oak fern

Matteuccia struthiopteris, ostrich fern

Onoclea sensibilis, sensitive fern

Osmunda cinnamomea, cinnamon fern

Osmunda claytoniana, interrupted fern

Osmunda regalis, royal fern

Pellaea atropupurea, purple cliffbrake

Phegopteris hexagonoptera, broad beech fern

Polypodium virginianum, rock polypody

Polystichum acrostichoides, Christmas fern

Polystichum braunii, Braun's holly fern

Thelypteris noveboracensis, New York fern

Thelypteris palustris, marsh fern

Woodsia obtusa, blunt-lobed woodsia

Woodwardia areolata, netted chain fern

Woodwardia virginica, chain fern

MID-ATLANTIC AND MIDDLE STATES

New York, New Jersey, Pennsylvania, Maryland, Delaware, West Virginia, Virginia, Kentucky, Ohio, Indiana, Michigan

Shade and specimen trees

Acer rubrum, red maple
Acer saccharum, sugar maple
Acer negundo, box elder
Betula nigra, river birch
Carya glabra, pignut hickory
Carya laciniosa, shellbark hickory
Castanea dentata, American chestnut (blight-resistant variety)
Fagus grandifolia, American beech
Fraxinus americana, white ash
Fraxinus pennsylvanica, green ash
Liquidambar styraciflua, sweet gum
Liriodendron tulipifera, tulip tree
Nyssa sylvatica, black gum
Platanus occidentalis, American sycamore
Quercus alba, white oak
Quercus coccinia, scarlet oak
Quercus falcata, southern red oak
Quercus palustris, pin oak
Quercus phellos, willow oak
Quercus rubra, red oak
Quercus velutina, black oak
Tilia americana, basswood
Ulmus americana, American elm (blight-resistant variety)

Shrubs and understory trees

Acer pensylvanicum, striped maple
Acer spicatum, mountain maple
Alnus serrulata, smooth alder
Amelanchier arborea, downy serviceberry, shadblow serviceberry
Amelanchier canadensis, shadbush
Amelanchier laevis, smooth serviceberry
Amorpha fruticosa, false indigo
Aralia spinosa, devil's walking stick
Aronia arbutifolia, red chokeberry
Aronia melanocarpa, black chokeberry
Asimina triloba, pawpaw
Baccharis halimifolia, groundsel bush
Carpinus caroliniana, ironwood
Castanea pumila, chinquapin
Ceanothus americanus, New Jersey tea
Celtis occidentalis, common hackberry
Cephalanthus occidentalis, buttonbush
Cercis canadensis, redbud
Chionanthus virginicus, fringe tree
Clethra alnifolia, sweet pepper bush
Comptonia peregrina, sweet fern
Cornus alternifolia, alternate-leaf dogwood
Cornus amomum, silky dogwood
Cornus florida, flowering dogwood
Cornus racemosa, gray dogwood
Cornus sericea, redtwig dogwood

Corylus americana, American hazelnut
Dirca palustris, leatherwood
Euonymus americanus, American strawberry bush
Hamamelis virginiana, witch hazel
Hydrangea arborescens, smooth hydrangea
Ilex decidua, possum haw
Ilex glabra, inkberry
Ilex opaca, American holly
Ilex verticillata, winterberry
Itea virginica, Virginia sweetspire
Kalmia angustifolia, sheep laurel
Kalmia latifolia, mountain laurel
Lindera benzoin, spicebush
Magnolia virginiana, sweetbay magnolia
Marella pensylvanica, northern bayberry
Ostrya virginiana, American hop hornbeam
Rhododendron arborescens, sweet azalea
Rhododendron atlanticum, coastal azalea
Rhododendron calendulaceum, flame azalea
Rhododendron maximum, great laurel
Rhododendron periclymenoides, pink azalea, pinxter azalea
Rhododendron prinophyllum, roseshell azalea
Rhododendron viscosum, swamp azalea
Rhus copallina, winged sumac
Rhus glabra, smooth sumac
Rhus typhina, staghorn sumac
Rosa blanda, meadow rose
Rosa carolina, Carolina rose
Rosa palustris, swamp rose
Rosa setigera, prairie rose
Rosa virginiana, Virginia rose
Salix discolor, pussy willow
Salix nigra, black willow
Salix sericea, silky willow
Sambucus canadensis, elderberry
Sassafras albidum, sassafras
Spiraea alba, white meadowsweet
Spiraea tomentosa, steeplebush
Ulmus rubra, slippery elm
Vaccinium angustifolium, lowbush blueberry
Vaccinium corymbosum, highbush blueberry
Viburnum acerifolium, mapleleaf viburnum
Viburnum dentatum, arrowwood
Viburnum nudum, smooth witherod
Viburnum prunifolium, blackhaw

Conifers

Chamaecyparis thyoides, Atlantic white cedar
Juniperus virginiana, eastern red cedar
Pinus strobus, white pine
Pinus virginiana, Virginia pine
Tsuga canadensis, eastern hemlock

Vines

Bignonia capreolata, cross vine
Campsis radicans, trumpet vine
Clematis virginiana, virgin's bower

Lonicera sempervirens, coral honeysuckle

Parthenocissus quinquefolia, Virginia creeper

Wisteria frutescens, American wisteria

Streamside plants

Acer negundo, box elder

Acer rubrum, red maple

Alnus serrulata, smooth alder

Aronia arbutifolia, red chokeberry

Asimina triloba, pawpaw

Baccharis halimifolia, groundsel bush

Betula nigra, river birch

Celtis occidentalis, common hackberry

Cephalanthus occidentalis, buttonbush

Chamaedaphne calyculata, leatherleaf

Clethra alnifolia, sweet pepper bush

Cornus amomum, silky dogwood

Cornus sericea, redtwig dogwood

Fraxinus pennsylvanica, green ash

Liquidambar styraciflua, sweet gum

Magnolia virginiana, sweetbay magnolia

Marella pensylvanica, northern bayberry

Nyssa sylvatica, black gum

Platanus occidentalis, American sycamore

Populus deltoides, eastern cottonwood

Quercus bicolor, swamp white oak

Quercus palustris, pin oak

Quercus phellos, willow oak

Rhododendron viscosum, swamp azalea

Rosa palustris, swamp rose

Salix discolor, pussy willow

Salix nigra, black willow

Salix sericea, silky willow

Sambucus canadensis, elderberry

Spiraea tomentosa, steeplebush

Ground covers

Antennaria dioica, pussytoes

Arctostaphylos uva-ursi, bearberry

Asarum canadense, wild ginger

Carex glaucodea, blue wood sedge

Carex pensylvanica, Pennsylvania sedge

Carex plantaginea, plantainleaf sedge

Carex platyphylla, broadleaf sedge

Carex stricta, tussock sedge

Chrysogonum virginianum, green and gold

Fragaria virginiana, common strawberry

Gaultheria procumbens, teaberry, wintergreen

Hepatica americana, round-lobed hepatica

Maianthemum canadense, Canada mayflower

Mitchella repens, partridgeberry

Pachysandra procumbens, Allegheny spurge

Phlox divaricata, wild blue phlox

Phlox maculata, wild sweet William

Phlox stolonifera, creeping phlox

Podophyllum peltatum, mayapple

Potentilla canadensis, dwarf cinquefoil
Potentilla simplex, common cinquefoil
Sedum ternatum, mountain stonecrop
Tiarella cordifolia, eastern foamflower
Viola blanda, sweet white violet
Viola pallens, northern white violet
Viola papilionacea, common blue violet
Viola pedata, birdsfoot violet
Viola septentrionalis, northern blue violet
Waldsteinia fragarioides, barren strawberry

Herbaceous perennials, dry sites

Agrimonia parviflora, small agrimony
Amsonia tabernaemontana, willowleaf bluestar
Anemone canadensis, Canada anemone
Aquilegia canadensis, wild columbine
Arisaema triphyllum, Jack-in-the-pulpit
Asclepias tuberosa, butterfly weed
Aster divaricatus, white wood aster
Aster lateriflorus, calico aster
Aster novae-angliae, New England aster
Aster oblongifolius, aromatic aster
Aster spectabilis, seaside aster
Baptisia australis, blue false indigo
Cardamine concatenata, cutleaf toothwort
Chrysopis mariana, Maryland golden aster
Cimicifuga racemosa, black cohosh
Coreopsis lanceolata, tickseed
Coreopsis verticillata, whorled coreopsis
Eupatorium hyssopifolium, hyssopleaf thoroughwort
Eupatorium perfoliatum, boneset
Eupatorium rotundifolium, round-leaved boneset
Eupatorium serotinum, late-blooming thoroughwort
Euphorbia corollata, flowering spurge
Geranium maculatum, wild geranium
Heliopsis helianthoides, oxeye
Heuchera americana, alumroot, coral bells
Lespedeza virginica, slender bushclover
Mertensia virginica, Virginia bluebells
Oenothera fruticosa, sundrops
Packera aurea, golden ragwort
Phlox maculata, meadow phlox
Pycnanthemum muticum, short-toothed mountain mint
Pycnanthemum tenuifolium, narrow-leaved mountain mint
Ratibida pinnata, pinnate prairie coneflower
Rudbeckia fulgida, orange coneflower
Rudbeckia nitida, shining coneflower
Salvia lyrata, lyre-leaved sage

Sedum ternatum, mountain stonecrop

Senna hebecarpa, wild senna

Silene virginica, fire pink

Sisyrinchium angustifolium, blue-eyed grass

Solidago caesia, blue-stem goldenrod

Solidago canadensis, Canada goldenrod

Solidago graminifolia, grass-leaved goldenrod

Solidago rugosa, rough-stem goldenrod

Solidago speciosa, showy goldenrod

Veronicastrum virginicum, Culver's root

Herbaceous perennials, moist sites

Aconitum uncinatum, eastern monkshood

Actaea pachypoda, white baneberry

Agastache scrophulariifolia, giant purple hyssop

Aruncus dioicus, goat's beard

Asclepias incarnata, swamp milkweed

Aster puniceus, purple-stem aster

Astilbe biternata, American astilbe

Boltonia asteroides, false aster, white doll's daisy

Chelone glabra, turtlehead

Coreopsis rosea, pink coreopsis

Dicentra cucullaria, Dutchman's breeches

Dicentra eximia, fringed bleeding heart

Erythronium americanum, yellow trout lily

Eupatorium coelestinum, mistflower

Eupatorium dubium, common Joe-Pye weed

Eupatorium fistulosum, hollowstem Joe-Pye weed

Eupatorium rugosum, white snakeroot

Filipendula rubra, queen of the prairie

Helenium autumnale, Helen's flower, sneezeweed

Helianthus angustifolius, swamp sunflower

Hibiscus moscheutos, rose mallow

Iris versicolor, blue flag iris

Jeffersonia diphylla, twinleaf

Liatris spicata, spiked blazing star

Lobelia cardinalis, cardinal flower

Lobelia siphilitica, great blue lobelia

Ludwigia alternifolia, seedbox

Lysimachia ciliata, yellow loosestrife

Mimulus ringens, Allegheny monkey flower

Monarda didyma, beebalm

Penstemon digitalis, foxglove beardtongue

Phlox paniculata, summer phlox

Polemonium reptans, creeping Jacob's ladder

Polygonatum commutatum, Solomon's seal

Rudbeckia laciniata, cutleaf coneflower

Verbena hastata, blue vervain

Vernonia noveboracensis, New York ironweed

Grasses, sedges, and rushes

Andropogon gerardii, big bluestem
Andropogon ternarius, splitbeard bluestem
Andropogon virginicus, broomsedge
Carex flaccosperma, thinfruit sedge
Carex pensylvanica, Pennsylvania sedge
Carex vulpinoidea, fox sedge
Chasmanthium latifolium, river oats
Elymus canadensis, Canada wild rye
Elymus hystrix, bottlebrush grass
Elymus virginicus, Virginia wild rye
Eragrostis spectabilis, purple lovegrass
Juncus effusus, common rush
Panicum virgatum, switchgrass
Saccharum giganteum, giant plume grass
Schizachyrium scoparium, little bluestem
Sorghastrum nutans, Indiangrass
Tripsacum dactyloides, eastern gamagrass

Ferns

Adiantum pedatum, maidenhair fern
Athyrium filix-femina, lady fern
Dennstaedtia punctilobula, hayscented fern
Onoclea sensibilis, sensitive fern
Osmunda cinnamomea, cinnamon fern
Osmunda regalis, royal fern
Polystichum acrostichoides, Christmas fern
Thelypteris noveboracensis, New York fern

Native plants relatively unpalatable to white-tailed deer

Trees, shrubs, and vines

Acer rubrum, red maple
Amelanchier species, serviceberry
Amorpha fruticosa, false indigo
Asclepias species, milkweeds
Betula species, birches
Dirca palustris, leatherwood
Juglans nigra, black walnut
Lindera benzoin, spicebush
Liquidambar styraciflua, sweet gum
Lonicera sempervirens, coral honeysuckle
Magnolia virginiana, sweetbay magnolia
Marella pensylvanica, northern bayberry
Platanus occidentalis, American sycamore
Prunus serotina, black cherry
Salix nigra, black willow
Sassafras albidum, sassafras

Herbaceous perennials

Aconitum uncinatum, eastern monkshood
Actaea pachypoda, white baneberry
Agastache scrophulariifolia, giant purple hyssop

Agrimonia parviflora, small agrimony
Aquilegia canadensis, wild columbine
Arisaema triphyllum, Jack-in-the-pulpit
Aruncus dioicus, goat's beard
Asarum canadense, wild ginger
Asclepias species, milkweeds
Aster oblongifolius, aromatic aster
Baptisia australis, blue false indigo
Clematis virginiana, virgin's bower
Coreopsis lanceolata, tickseed
Coreopsis rosea, pink coreopsis
Dicentra eximia, fringed bleeding heart
Euphorbia corollata, flowering spurge
Geranium maculatum, wild geranium
Helenium autumnale, Helen's flower, sneezeweed
Hibiscus moscheutos, rose mallow
Iris versicolor, blue flag iris
Jeffersonia diphylla, twinleaf
Liatris spicata, spiked blazing star
Mimulus ringens, Allegheny monkey flower
Monarda fistulosa, wild bergamot
Penstemon digitalis, foxglove beardtongue
Symplocarpus foetidus, skunk cabbage

SOUTHEAST

Louisiana, Arkansas, Alabama, Mississippi, Florida, Georgia, South Carolina, North Carolina, Tennessee

Shade and specimen trees

Acer rubrum, red maple
Acer saccharinum, silver maple
Betula lenta, black birch
Betula nigra, river birch
Carya illinoensis, pecan
Carya ovata, shagbark hickory
Catalpa bignonioides, catalpa
Cladrastis kentukea, yellowwood
Diospyros virginiana, persimmon
Fagus grandifolia, American beech
Fraxinus americana, white ash
Gleditsia triacanthos, honey locust
Gymnocladus dioicus, Kentucky coffee tree
Juglans nigra, black walnut
Liquidambar styraciflua, sweet gum
Liriodendron tulipifera, tulip tree
Magnolia acuminata, cucumber tree
Magnolia grandiflora, southern magnolia
Magnolia macrophylla, large-leaved umbrella tree
Nyssa sylvatica, black gum
Oxydendron arboreum, sourwood
Prunus serotina, black cherry
Quercus alba, white oak
Quercus coccinea, scarlet oak
Quercus palustris, pin oak
Quercus phellos, willow oak
Quercus virginiana, live oak
Robinia pseudoacacia, black locust

Sapindus drummondii, soapberry
Tilia americana var. *heterophylla*, basswood

Shrubs and understory trees

Acer negundo, box elder
Aesculus pavia, red buckeye
Amelanchier arborea, downy serviceberry, shadblow serviceberry
Aronia melanocarpa, black chokeberry
Asimina triloba, pawpaw
Baccharis halimifolia, groundsel bush
Callicarpa americana, beautyberry
Calycanthus floridus, Carolina allspice
Carpinus caroliniana, ironwood
Ceanothus americanus, New Jersey tea
Celtis laevigata, Mississippi hackberry
Cercis canadensis, redbud
Chionanthus virginicus, fringe tree
Clethra alnifolia, sweet pepper bush
Comptonia peregrina, sweet fern
Cornus alternifolia, alternate-leaf dogwood
Cornus florida, flowering dogwood
Cornus stolonifera, red osier dogwood
Corylus americana, American hazelnut
Cotinus obovatus, smoke tree
Crataegus phaenopyrum, Washington hawthorn
Crataegus viridis, green hawthorn
Cyrilla racemiflora, titi
Diervilla sessilifolia, southern bush honeysuckle
Euonymus americanus, American strawberry bush
Fothergilla gardenii, dwarf fothergilla
Franklinia alatamaha, Franklin tree
Halesia carolina, Carolina silverbell
Hamamelis virginiana, witch hazel
Hydrangea arborescens, smooth hydrangea
Hydrangea quercifolia, oakleaf hydrangea
Hypericum densiflorum, bushy St. John' s wort
Ilex decidua, possum haw
Ilex glabra, inkberry
Ilex opaca, American holly
Ilex verticillata, winterberry
Illicium floridanum, Florida anise
Itea virginica, Virginia sweetspire
Kalmia latifolia, mountain laurel
Leiophyllum buxifolium, sand myrtle
Lindera benzoin, spicebush
Maclura pomifera, Osage orange
Magnolia virginiana, sweetbay magnolia
Malus angustifolia, southern crabapple
Morus rubra, red mulberry
Neviusia alabamensis, snow wreath
Osmanthus americanus, devilwood
Ostrya virginiana, American hop hornbeam
Persea borbonia, red bay
Physocarpus opulifolius, common ninebark

Pieris floribunda, fetterbush
Prunus caroliniana, cherry laurel
Ptelea trifoliata, hoptree
Rhododendron atlanticum, coastal azalea
Rhododendron calendulaceum, flame azalea
Rhododendron canescens, wild azalea
Rhododendron catawbiense, Catawba rosebay
Rhododendron maximum, great laurel
Rhododendron periclymenoides, pink azalea, pinxter azalea
Rhododendron viscosum, swamp azalea
Rhus aromatica, fragrant sumac
Robinia hispida, bristly locust
Rosa carolina, Carolina rose
Rosa virginiana, Virginia rose
Salix nigra, black willow
Sambucus canadensis, elderberry
Sassafras albidum, sassafras
Sorbus americana, American mountain ash
Staphylea trifoliata, bladdernut
Stewartia ovata, mountain camellia
Styrax americana, American snowbell
Symphoricarpos orbiculatus, coralberry
Vaccinium ashei, rabbiteye blueberry
Vaccinium corymbosum, highbush blueberry
Viburnum acerifolium, mapleleaf viburnum
Viburnum dentatum, arrowwood
Viburnum nudum, smooth witherod
Viburnum prunifolium, blackhaw
Zenobia pulverulenta, honeycup

Conifers

Chamaecyparis thyoides, Atlantic white cedar
Juniperus virginiana, eastern red cedar
Pinus palustris, longleaf pine
Pinus taeda, loblolly pine
Pinus virginiana, Virginia pine
Taxus floridana, Florida yew
Tsuga caroliniana, Carolina hemlock

Vines

Aristolochia durior, Dutchman's pipe
Bignonia capreolata, cross vine
Campsis radicans, trumpet vine
Clematis virginiana, virgin's bower
Gelsemium sempervirens, Carolina jessamine
Lonicera flava, yellow honeysuckle
Lonicera sempervirens, coral honeysuckle
Parthenocissus quinquefolia, Virginia creeper
Passiflora incarnata, passion vine
Schizophragma integrifolium, climbing hydrangea
Vitis rotundifolia, fox grape, muscadine
Wisteria frutescens, American wisteria

Streamside plants

Alnus serrulata, smooth alder
Cephalanthus occidentalis, buttonbush
Dirca palustris, leatherwood
Eubotrys racemosa, swamp doghobble
Fraxinus caroliniana, water ash
Gordonia lasianthus, loblolly bay
Leucothoe axillaris, coastal doghobble
Lyonia lucida, fetterbush
Pinckneya pubens, pinkneya
Taxodium distichum, bald cypress
Xanthorhiza simplicissima, yellowroot

Ground covers

Epigaea repens, trailing arbutus
Galax aphylla, galax
Gaylussacia brachycera, box huckleberry
Hexastylis arifolia, jug plant
Hypericum lloydii, sandhill St. John's wort
Mitchella repens, partridgeberry
Phlox subulata, moss pink
Shortia galacifolia, Oconee bells

Herbaceous perennials, dry sites

Amsonia ciliata, fringed bluestar
Asclepias tuberosa, butterfly weed
Aster dumosus, bushy aster
Baptisia alba, wild white indigo
Conradina canescens, false rosemary
Erythrina herbacea, coral bean
Gaillardia pulchella, blanketflower
Liatris elegans, pinkscale blazing star
Liatris spicata, spiked blazing star
Lilium catesbaei, pine lily
Lobelia cardinalis, cardinal flower
Monarda punctata, dotted horsemint
Penstemon multiflorus, manyflower beardtongue
Phlox pilosa, downy phlox
Rudbeckia hirta, black-eyed Susan
Ruellia caroliniensis, Carolina wild petunia
Salvia coccinea, scarlet sage
Sisyrinchium angustifolium, blue-eyed grass
Solidago sempervirens, seaside goldenrod
Tradescantia ohiensis, spiderwort
Vernonia angustifolia, tall ironweed

Herbaceous perennials, moist sites

Arisaema dracontium, green dragon
Asclepias perennis, aquatic milkweed
Crinum americanum, swamp lily
Eryngium yuccifolium, button snakeroot
Eupatorium coelestinum, mistflower
Helianthus angustifolius, swamp sunflower
Hibiscus moscheutos, rose mallow

Iris hexagona, prairie iris
Scutellaria integrifolia, rough skullcap
Teucrium canadense, wood sage
Viola lanceolata, bog white violet

Grasses, sedges, and rushes

Andropogon virginicus, broomsedge
Eragrostis spectabilis, purple lovegrass
Juncus effusus, common rush
Scirpus cyperinus, bulrush, woolgrass
Sorghastrum secundum, lopsided Indiangrass
Tripsacum dactyloides, eastern gamagrass

Ferns

Asplenium platyneuron, ebony spleenwort
Dryopteris ludoviciana, southern wood fern
Osmunda cinnamomea, cinnamon fern
Osmunda regalis, royal fern
Polypodium polypodioides, Resurrection fern
Polystichum acrostichoides, Christmas fern
Thelypteris kunthii, maiden fern
Woodwardia virginica, chain fern

MIDWEST AND EASTERN GREAT PLAINS

North Dakota, South Dakota, Nebraska, Kansas, Oklahoma, east Texas, Minnesota, Iowa, Missouri, Wisconsin, Illinois

Shade and specimen trees

Acer rubrum, red maple
Acer saccharum, sugar maple
Acer saccharinum, silver maple
Betula alleghaniensis, yellow birch
Betula nigra, river birch
Betula papyrifera, paper birch
Carya cordiformis, bitternut hickory
Carya ovata, shagbark hickory
Diospyros virginiana, persimmon
Fagus grandifolia, American beech
Fraxinus americana, white ash
Fraxinus nigra, black ash
Gleditsia triacanthos, honey locust
Gymnocladus dioicus, Kentucky coffee tree
Juglans cinerea, butternut
Juglans nigra, black walnut
Liriodendron tulipifera, tulip tree
Malus joensis, wild crabapple
Morus rubra, red mulberry
Nyssa sylvatica, black gum
Ostrya virginiana, American hop hornbeam
Platanus occidentalis, American sycamore
Populus deltoides, eastern cottonwood

Populus tremuloides, quaking aspen
Prunus serotina, black cherry
Quercus alba, white oak
Quercus macrocarpa, bur oak
Quercus muehlenbergii, chinkapin oak
Quercus palustris, pin oak
Quercus rubra, red oak
Quercus velutina, black oak
Robinia pseudoacacia, black locust
Sassafras albidum, sassafras
Tilia americana, basswood
Ulmus americana, American elm (blight-resistant variety)

Shrub and understory trees

Acer penslyvanicum, striped maple
Aesculus glabra, Ohio buckeye
Amelanchier arborea, downy serviceberry, shadblow serviceberry
Amelanchier humilis, low serviceberry
Amorpha canescens, lead plant
Amorpha fruticosa, false indigo
Aronia melanocarpa, black chokeberry
Asimina triloba, pawpaw
Atriplex canescens, four-winged saltbrush
Carpinus caroliniana, ironwood
Ceanothus americanus, New Jersey tea
Celtis occidentalis, common hackberry
Cephalanthus occidentalis, buttonbush
Cercis canadensis, redbud
Cornus alternifolia, alternate-leaf dogwood
Cornus amomum, silky dogwood
Cornus racemosa, gray dogwood
Cornus sericea, redtwig dogwood
Corylus americana, American hazelnut
Crataegus crus-galli, cockspur hawthorn
Crataegus mollis, downy hawthorn
Diervilla lonicera, bush honeysuckle
Dirca palustris, leatherwood
Euonymus atropurpureus, eastern wahoo, burning bush
Gaylussacia baccata, black huckleberry
Hamamelis virginiana, witchhazel
Ilex glabra, inkberry
Ilex opaca, American holly
Ilex verticillata, winterberry
Kalmia latifolia, mountain laurel
Lindera benzoin, spicebush
Myrica pensylvanica, northern bayberry
Nemopanthus mucronatus, American mountain holly
Ostrya virginiana, American hop hornbeam
Physocarpus opulifolius, common ninebark
Prunus americana, American plum
Prunus virginiana, chokecherry
Ptelea trifoliata, hoptree
Quercus prinoides, dwarf chinkapin oak
Rhamnus lanceolata, lanceleaf buckthorn
Rhus aromatica, fragrant sumac
Rhus copallina, winged sumac

Rhus glabra, smooth sumac
Rhus typhina, staghorn sumac
Ribes americanum, wild black currant
Ribes missouriense, Missouri gooseberry
Rosa acicularis, prickly rose
Rosa arkansana, Arkansas rose
Rosa blanda, meadow rose
Rosa setigera, prairie rose
Rubus flagellaris, northern dewberry
Rubus idaeus, red raspberry
Rubus occidentalis, black raspberry
Salix petiolaris, meadow willow
Sambucus canadensis, elderberry
Shepherdia argentea, sliver buffaloberry
Shepherdia canadensis, russet buffaloberry
Sorbus americana, mountain ash
Spiraea alba, white meadowsweet
Spiraea tomentosa, steeplebush
Staphylea trifoliata, bladdernut
Symphoricarpos albus, snowberry
Symphoricarpos orbiculatus, coralberry
Vaccinium angustifolium, low sweet blueberry
Vaccinium corymbosum, highbush blueberry
Vaccinium vitis-idaea, mountain cranberry
Viburnum acerifolium, mapleleaf viburnum
Viburnum dentatum, arrowwood
Viburnum lentago, nannyberry
Viburnum nudum, smooth witherod, possumhaw
Viburnum prunifolium, blackhaw
Zanthoxylum americanum, pricklyash

Conifers

Abies balsamea, balsam fir
Juniperus communis, common juniper
Juniperus virginiana, red cedar
Larix laricna, tamarack
Picea glauca, white spruce
Picea mariana, black spruce
Pinus banksiana, Jack pine
Pinus resinosa, red pine
Pinus rigida, pitch pine
Pinus strobus, white pine
Taxus canadensis, Canada yew
Thuja occidentalis, northern white-cedar

Vines

Ampelopsis cordata, raccoon grape
Celastrus scandens, American bittersweet
Lonicera dioica, limber honeysuckle
Menispermum canadense, moonseed
Parthenocissus quinquefolia, Virginia creeper
Parthenocissus vitacea, woodbine
Vitis cinerea, winter grape
Vitis riparia, riverbank grape
Vitis vulpina, frost grape

Streamside plants

Alnus incana, gray alder
Alnus viridis, mountain alder
Betula pumila, bog birch
Cephalanthus occidentalis, buttonbush
Chamaedaphne calyculata, leatherleaf
Cornus amomum, silky dogwood
Dirca palustris, leatherwood
Ilex verticillata, winterberry
Salix amygdaloides, peachleaf willow
Salix discolor, pussy willow
Salix eriocephala, Missouri River willow
Salix humilis, prairie willow
Salix interior, sandbar willow
Salix lutea, yellow willow
Salix nigra, black willow

Ground covers

Asarum canadense, wild ginger
Cornus canadensis, bunchberry
Gaultheria procumbens, teaberry
Gaylussacia brachycera, box huckleberry
Jeffersonia diphylla, twinleaf
Juniperus horizontalis, horizontal juniper
Maianthemum canadense, Canada mayflower
Mitchella repens, partridgeberry
Mahonia repens, creeping mahonia
Phlox subulata, moss pink
Podophyllum peltatum, mayapple
Potentilla tridentata, three-toothed cinquefoil
Sibbaldiopsis tridentata, shrubby fivefingers
Tiarella cordifolia, eastern foamflower
Vaccinium vitis-idaea, mountain cranberry
Waldsteinia fragarioides, barren strawberry

Herbaceous perennials, dry sites

Anemone canadensis, Canada anemone
Aquilegia canadensis, wild columbine
Asclepias syriaca, common milkweed
Asclepias tuberosa, butterfly weed
Aster ericoides, white heath aster
Baptisia alba, wild white indigo
Baptisia australis, blue false indigo
Campanula rotundifolia, harebell
Caulophyllum thalictroides, blue cohosh
Cimicifuga racemosa, black cohosh
Coreopsis lanceolata, tickseed
Cypripedium parviflorum, lesser yellow lady's slipper
Echinacea purpurea, purple coneflower
Eupatorium rugosum, white snakeroot
Geranium maculatum, wild geranium
Geum triflorum, prairie smoke
Hedyotis caerulea, bluets
Heliopsis helianthoides, oxeye
Hepatica acutiloba, sharp-lobed hepatica

Heuchera americana, alumroot, coral bells
Hydrastis canadensis, goldenseal
Iris cristata, crested iris
Lilium philadelphicum, wood lily
Lupinus perennis, blue lupine
Mitella diphylla, miterwort
Oenothera fruticosa, sundrops
Opuntia humifusa, eastern prickly pear
Panax quinquefolius, ginseng
Ratibida pinnata, pinnate prairie coneflower
Rudbeckia fulgida, orange coneflower
Rudbeckia hirta, black-eyed Susan
Rudbeckia laciniata, cutleaf coneflower
Sanguinaria canadensis, bloodroot
Sisyrinchium angustifolium, blue-eyed grass
Solidago caesia, blue-stem goldenrod
Soldago rugosa, rough-stem goldenrod
Solidago speciosa, showy goldenrod
Tradescantia ohiensis, spiderwort
Trillium flexipes, bent trillium
Trillium grandiflorum, white trillium
Trillium sessile, red trillium
Uvularia grandiflora, large-flowered bellwort
Viola pedata, bird's foot violet

Herbaceous perennials, moist and wet sites

Actaea rubra, red baneberry
Anemone quinquefolia, wood anemone
Arisaema triphyllum, Jack-in-the-pulpit
Asclepias incarnata, swamp milkweed
Aster novae-angliae, New England aster
Boltonia asteroides, false aster, white doll's daisy
Calla palustris, water arum
Caltha palustris, marsh marigold
Chelone glabra, turtlehead
Cimicifuga racemosa, black cohosh
Clintonia borealis, blue bead lily
Coreopsis rosea, pink coreopsis
Delphinium tricorne, dwarf larkspur
Dicentra canadensis, squirrel corn
Dicentra cucullaria, Dutchman's breeches
Dodecantheon meadia, shooting star
Erythronium americanum, yellow trout lily
Eupatorium maculatum, spotted Joe-Pye weed
Eupatorium perfoliatum, boneset
Filipendula rubra, queen of the pairie
Gentiana clausa, closed gentian
Helenium autumnale, Helen's flower, sneezeweed
Hepatica americana, round-lobed hepatica
Hibiscus moscheutos, rose mallow
Iris versicolor, blue flag iris

Liatris spicata, spiked blazing star
Lobelia cardinalis, cardinal flower
Lobelia siphilitica, great blue lobelia
Lupinus perennis, blue lupine
Mertensia virginica, Virginia bluebells
Monarda didyma, beebalm
Monarda fistulosa, wild bergamot
Packera aurea, golden ragwort
Penstemon digitalis, foxglove beardtongue
Penstemon hirsutus, hairy beardtongue
Phlox divaricata, wild blue phlox
Phlox maculata, wild sweet William
Physostegia virginiana, false dragonhead
Polemonium reptans, creeping Jacob's ladder
Polygonatum commutatum, Solomon's seal
Smilacina racemosa, false Solomon's seal
Stylophorum diphyllum, celandine poppy
Teucrium canadense, wood sage
Thalictrum dasycarpum, purple meadow rue
Thalictrum pubescens, tall meadow rue
Tradescantia virginiana, Virginia spiderwort
Trientalis borealis, starflower
Veronicastrum virginicum, Culver's root
Viola cucullata, marsh blue violet
Viola pubescens, yellow violet
Viola sororia, hooded violet
Zizia aurea, golden Alexanders

Grasses, sedges, and rushes

Acorus calamus, sweetflag
Andropogon gerardii, big bluestem
Andropogon virginicus, broomsedge
Aristida purpurea, purple three awn
Bouteloua gracilis, blue gramagrass
Buchloe dactyloides, buffalograss
Carex muskingumensis, palm sedge
Carex pensylvanica, Pennsylvania sedge
Carex plantaginea, plantain sedge
Carex platyphylla, broadleaf sedge
Chasmanthium latifloium, river oats
Deschampsia cespitosa, tufted hairgrass
Elymus glaucus, blue wild rye
Eragrostis spectabilis, purple lovegrass
Juncus effusus, common rush
Panicum virgatum, switchgrass
Scirpus cyperinus, bulrush, woolgrass
Schizacharium scoparium, little bluestem
Sorghastrum nutans, Indian grass
Spartina pectinata, prairie cord-grass
Sporobolus heterolepis, prairie dropseed
Tridens flavus, purpletop
Typha angustifolia, narrow-leaf cattail

Ferns

Adiantum pedatum, maidenhair fern
Asplenium trichomanes, spleenwort, maidenhair spleenwort
Athyrium filix-femina, lady fern
Botrychium virginianum, rattlesnake fern
Cheilanthes lanosa, hairy lip fern
Cryptogramma stelleri, slender rock brake
Cystopteris fragilis, brittle bladder fern, fragile fern
Dennstaedtia punctilobula, hayscented fern
Dryopteris cristata, crested wood fern
Dryopteris filix-mas, male fern
Dryopteris goldiana, Goldie's fern
Dryopteris intermedia, evergreen wood fern
Dryopteris marginalis, marginal shield fern
Gymnocarpium dryopteris, oak fern
Matteuccia struthiopteris, ostrich fern
Onoclea sensibilis, sensitive fern
Osmunda cinnamomea, cinnamon fern
Osmunda claytoniana, interrupted fern
Osmunda regalis, royal fern
Pellaea atropupurea, purple cliffbrake
Phegopteris hexagonoptera, broad beech fern
Polypodium virginianum, rock polypody
Polystichum acrostichoides, Christmas fern
Polystichum braunii, Braun's holly fern
Pteridium aquilinum, bracken fern
Thelypteris palustris, marsh fern
Woodsia obtusa, blunt-lobed woodsia
Woodwardia virginica, chain fern

PACIFIC NORTHWEST

Western Washington, western Oregon, northern California

Shade and specimen trees

Betula occidentalis, water birch
Betula papyrifera, paper birch
Chrysolepis chrysophylla, golden chinquapin
Fraxinus latifolia, Oregon ash
Lithocarpus densiflorus, tanbark oak
Populus trichocarpa, black cottonwood
Quercus chrysolepis, canyon live oak
Quercus garryana, Garry oak
Quercus kelloggii, California black oak
Umbellularia californica, California bay laurel

Shrubs and understory trees

Acer circinatum, vine maple
Acer macrophyllum, bigleaf maple
Amelanchier alnifolia, western serviceberry
Arbutus menziesii, Pacific madrone
Arctostaphylos columbiana, hairy manzanita
Artemisia tridentata, sagebrush
Baccharis pilularis, coyotebrush
Ceanothus integerrimus, deerbrush
Ceanothus thyrsiflorus, blueblossum
Cercocarpus montanus, mountain mahogany
Chrysolepis sempervirens, bush chinquapin
Chrysothamnus nauseosus, rabbitbrush
Cornus nuttallii, Pacific dogwood
Corylus cornuta, beaked hazel
Euonymus occidentalis, western wahoo
Garrya elliptica, silk-tassel bush
Gaultheria shallon, salal
Grindelia integrifolia, resinweed
Holodiscus discolor, oceanspray
Kalmiopsis leachiana, kalmiopsis
Ledum groenlandicum, Labrador tea
Lithocarpus densiflorus var. *echinoides*, shrub tanbark oak
Mahonia aquifolium, Oregon grape
Mahonia nervosa, low Oregon grape
Mimulus aurantiacus, shrubby monkey flower
Myrica californica, California wax myrtle
Paxistima myrsinites, Oregon box
Penstemon fruticosus, shrubby penstemon
Philadelphus lewisii, mock orange
Phyllodoce empetriformis, mountain heather
Populus tremuloides, quaking aspen
Potentilla fruticosa, shrubby cinquefoil
Quercus vaccinifolia, huckleberry oak
Rhamnus californica, coffeeberry
Rhododendron macrophyllum, Pacific rhododendron
Rhus glabra, smooth sumac
Ribes sanguineum, red-flowering currant
Rosa nutkana, Nootka rose
Rubus spectabilis, salmonberry
Sambucus cerulea, blue elderberry
Shepherdia canadensis, russet buffaloberry
Sorbus sitchensis, Sitka mountain ash
Spiraea douglasii, rose spirea
Symphoricarpos albus, snowberry
Vaccinium ovatum, evergreen huckleberry
Vaccinium parvifolium, red huckleberry
Viburnum edule, moosewood viburnum

Conifers

Abies amabilis, Pacific silver fir
Abies concolor, white fir
Abies procera, noble fir
Calocedrus decurrens, incense cedar

Chamaecyparis lawsoniana, Port Orford cedar
Chamaecyparis nootkatensis, yellow cedar
Juniperus occidentalis, western juniper
Juniperus scopulorum, Rocky Mountain juniper
Larix occidentalis, tamarack
Picea engelmannii, Engelmann spruce
Picea sitchensis, Sitka spruce
Pinus banksiana, Jack pine
Pinus contorta, lodgepole pine
Pinus flexilis, limber pine
Pinus lambertiana, sugar pine
Pinus ponderosa, ponderosa pine
Pseudotsuga menziesii, Douglas fir
Sequoiadendron giganteum, giant sequoia
Taxus brevifolia, western yew
Thuja plicata, western red cedar
Tsuga heterophylla, western hemlock

Vines

Clematis columbiana, rock clematis
Lonicera ciliosa, trumpet honeysuckle
Lonicera hispidula, hairy honeysuckle
Vitis californica, western wild grape

Streamside plants

Alnus rubra, red alder
Andromeda polifolia, bog rosemary
Betula glandulosa, swamp birch
Cornus stolonifera, red osier dogwood
Kalmia occidentalis, bog laurel
Kalmia polifolia, swamp laurel
Ledum glandulosum, trapper's tea
Ledum groenlandicum, Labrador tea
Leucothoe davisiae, western leucothoe
Lonicera involucrata, twinberry
Myrica gale, sweetgale
Physocarpus capitatus, Pacific ninebark
Rhododendron occidentale, western azalea
Salix scouleriana, Scouler's willow
Vaccinium oxycoccus, cranberry

Ground covers

Achlys triphylla, vanilla leaf
Arctostaphylos uva-ursi, bearberry
Asarum caudatum, long-tailed wild ginger
Ceanothus prostratus, Mahala mat
Ceanothus pumilus, Siskiyou mat
Cornus canadensis, bunchberry
Dicentra formosa, western bleeding heart
Empetrum nigrum, crowberry
Fragaria chiloensis, coastal strawberry
Gaultheria ovatifolia, slender wintergreen

Juniperus communis, common juniper

Linnaea borealis, twinflower

Loiseleuria procumbens, alpine azalea

Maianthemum dilatatum, false lily-of-the-valley

Oxalis oregana, redwood sorrel

Satureja douglasii, yerba buena

Selaginella oregana, Oregon spikemoss

Tiarella trifoliata, threeleaf foamflower

Trientalis arctica, northern starflower

Vancouveria hexandra, vancouveria

Herbaceous perennials, dry sites

Aquilegia formosa, red columbine

Aralia californica, western aralia

Aster foliaceus, leafy aster

Campanula rotundifolia, harebell

Clematis hirsutissima, hairy clematis

Corydalis scouleri, corydalis

Erythronium oreganum, fawn lily

Geranium viscosissimum, sticky purple geranium

Iris tenax, Oregon iris

Lilium columbianum, tiger lily

Lupinus polyphyllus, bigleaf lupine

Mertensia paniculata, tall bluebells

Oenothera biennis, evening primrose

Paeonia brownii, western peony

Penstemon acuminatus, sharpleaf penstemon

Scrophularia californica, figwort

Sisyrinchium bellum, western blue-eyed grass

Maianthemum racemosum, false Solomon's seal

Solidago canadensis, Canada goldenrod

Solidago graminifolia, lance-leaved goldenrod

Thermopsis montana, golden pea

Zigadenus elegans, mountain death camas

Herbaceous perennials, moist sites

Aconitum columbianum, Columbian monkshood

Aruncus dioicus, goat's beard

Geranium oreganum, Oregon geranium

Iliamna rivularis, streambank wild hollyhock

Mimulus guttatus, monkey flower

Peltiphyllum peltatum, umbrella plant

Primula parryi, Parry's primrose

Ferns

Adiantum pedatum, maidenhair fern

Asplenium trichomanes, spleenwort, maidenhair spleenwort

Athyrium filix-femina, lady fern

Blechnum spicant, deer fern

Cheilanthes gracillima, lace fern

Cryptogramma crispa, parsley fern

Cystopteris fragilis, brittle bladder fern, fragile fern

Gymnocarpium dryopteris, oak fern

Pellaea atropurpurea, purple cliffbrake

Polypodium glycyrrhiza, licorice fern

Polystichum munitum, sword fern

Woodwardia fimbriata, giant chain fern

SOUTHWEST

Southern California, Nevada, Arizona, New Mexico, west Texas

Shrubs and understory trees

Acacia farnesiana, opopanax

Acer grandidentatum, bigtooth maple

Amorpha canescens, lead plant

Arctostaphylos uva-ursi, bearberry

Artemisia cana, silver sage

Artemisia tridentata, sagebrush

Celtis reticulata, netleaf hackberry

Cercocarpus montanus, mountain mahogany

Chilopsis linearis, desert willow

Chrysothamnus nauseosus, rabbitbrush

Dendromecon rigida, bush poppy

Fendlera rupicola, cliff fendlerbush

Forestiera neomexicana, New Mexico privet

Fraxinus velutina, velvet ash

Heteromeles arbutifolia, Christmas berry

Holodiscus dumosus, cliff spirea

Larrea tridentata, creosote bush

Mahonia haematocarpa, desert holly

Philadelphus microphyllus, littleleaf mock orange

Potentilla fruticosa, shrubby cinquefoil

Prunus besseyi, western sand cherry

Purshia tridentata, antelope bitterbush

Rhus trilobata, skunkbush sumac

Ribes aureum, golden currant

Robinia neomexicana, New Mexico locust

Rosa stellata, desert rose

Shepherdia argentea, silver buffaloberry

Yucca baccata, broadleaf yucca

Yucca glauca, narrowleaf yucca

Conifers

Abies concolor, white fir

Calocedrus decurrens, incense cedar

Cupressus arizonica, Arizona cypress

Juniperus deppeana, alligator juniper

Juniperus monosperma, one-seed juniper

Picea pungens, blue spruce

Pinus aristata, bristlecone pine

Pinus contorta, lodgepole pine

Pinus flexilis, limber pine

Pinus ponderosa, ponderosa pine

Pinus strobiformis, southwest white pine

Pseudotsuga menziesii, Douglas fir

Vines

Clematis ligusticifolia, western virgin's bower

Clematis pseudoalpina, Rocky Mountain clematis

Maurandya antirrhiniflora, snapdragon vine

Vitis arizonica, canyon grape

Grasses

Andropogon gerardii, big bluestem

Aristida purpurea, purple three awn

Bouteloua curtipendula, sideoats grama

Bouteloua gracilis, blue gramagrass

Buchloe dactyloides, buffalograss

Eragrostis intermedia, plains lovegrass

Hierochloe odorata, sweetgrass

Panicum virgatum, switchgrass

Schizachyrium scoparium, little bluestem

Sorghastrum nutans, Indiangrass

Sporobolus wrightii, big sacaton

Stipa tenuissima, Mexican feathergrass

Herbaceous perennials

Abronia villosa, sand verbena

Agastache cana, wild hyssop

Allium cernuum, pink nodding onion

Aquilegia caerulea, Rocky Mountain columbine

Aquilegia chrysantha, golden-spurred columbine

Asclepias speciosa, showy milkweed

Asclepias tuberosa, butterfly weed

Aster bigelovii, purple aster

Baileya multiradiata, desert marigold

Berlandiera lyrata, chocolate flower

Callirhoe involucrata, wine cup

Castilleja integra, Indian paintbrush

Cleome serrulata, Rocky Mountain beeplant

Coreopsis tinctoria, Plains coreopsis

Dalea jamesii, James's dalea

Dichelostemma pulchellum, blue dicks

Echinacea purpurea, purple coneflower

Encelia farinosa, brittlebush

Eriogonum corymbosum, buckwheatbrush

Eriogonum umbellatum, sulphur-flower buckwheat

Erysimum asperum, western wallflower

Eschscholzia californica, California poppy

Gaillardia pulchella, blanketflower

Gilia aggregata, scarlet gilia

Helianthus maximiliani, Maximilian sunflower

Heuchera pulchella, mountain coral bells

Hymenoxys grandiflora, alpine sunflower

Ipomoea leptophylla, bush morning glory
Iris missouriensis, western blue flag
Liatris punctata, dotted blazing star
Liatris spicata, spiked blazing star
Lupinus argenteus, silvery lupine
Lupinus texensis, bluebonnet
Melampodium leucanthum, Blackfoot daisy
Mentzelia decapetala, ten-petal blazing star
Mirabilis multiflora, desert four o'clock
Monarda pectinata, horsemint
Oenothera missouriensis, Missouri evening primrose
Oenothera speciosa, Mexican evening primrose
Penstemon barbatus, scarlet bugler
Penstemon grandiflorus, large-flowered penstemon
Penstemon secundiflorus, sidebells penstemon
Phacelia campanularia, California bluebells
Polemonium foliosissimum, Jacob's ladder
Ratibida columnifera, Mexican hat
Rudbeckia hirta, black-eyed Susan
Salvia azurea, blue sage
Silene laciniata, Indian pink
Sphaeralcea ambigua, desert mallow
Tradescantia occidentalis, western spiderwort
Zauschneria latifolia, hummingbird trumpet
Zinnia grandiflora, prairie zinnia

APPENDIX TWO

Host Plants of Butterflies and Showy Moths

BUTTERFLIES	HOST PLANT
Blues (Lycaenidae)	
Eastern tailed blue (*Everes comyntas*)	Legume family (Fabaceae)
Spring azure (*Celastrina ladon*)	Dogwoods (*Cornus*), New Jersey tea (*Ceanothus americanus*), Viburnums (*Viburnum*)
Brush-footed Butterflies (Nymphalidae)	
American lady (*Vanessa virginiensis*)	Composite family (Asteraceae), Pearly everlasting (*Anaphalis margaritacea*), Pussytoes (*Antennaria dioica*), Sweet everlasting (*Gnaphalium obtusifolium*)
Appalachian brown (*Satyrodes appalachia*)	Georgia bulrush (*Scirpus georgianus*), Graceful sedge (*Carex gracillima*), Hairy sedge (*Carex lacustris*), Narrowfruit horned beaksedge (*Rhynchospora inundata*), Upright sedge (*Carex stricta*), Woolly sedge (*Carex lanuginosa*)

(continued)

BUTTERFLIES	HOST PLANT
Common buckeye (*Junonia coenia*)	Blue toadflax (*Linaria canadensis*)
Common wood nymph (*Cercyonis pegala*)	Alkali grass (*Puccinellia nuttalliana*), Big bluestem (*Andropogon gerardii*), Needlegrass (*Stipa spartea*), Redtop (*Tridens flavus*)
Eastern comma (*Polygonia comma*)	Elm family (Ulmaceae), Hackberries (*Celtis*), Nettles (*Urtica*)
Great spangled fritillary (*Speyeria cybele*)	Violets (*Viola*)
Hackberry butterfly (*Asterocampa celtis*)	Common hackberry (*Celtis occidentalis*)
Little wood satyr (*Megisto cymela*)	Centipede grass (*Eremochloa ophiuroides*)
Meadow fritillary (*Boloria bellona*)	Violets (*Viola*)
Monarch (*Danaus plexippus*)	Milkweeds, butterfly weeds (*Asclepias*)
Mourning cloak (*Nymphalis antiopa*)	Birches (*Betula*), Elms (*Ulmus*), Hackberries (*Celtis*), Nettles (*Urtica*), Poplars (*Populus*), Willows (*Salix*)
Painted lady (*Vanessa cardui*)	Pearly everlasting (*Anaphalis margaritacea*), Sweet everlasting (*Gnaphalium obtusifolium*)
Pearl crescent (*Phyciodes tharos*)	Aster family (Asteraceae)
Question mark (*Polygonia interrogationis*)	Elms (*Ulmus*), Nettles (*Urtica*)
Red-spotted purple (*Limenitis arthemis astyanax*)	Cherries (*Prunus*), Willows (*Salix*)

BUTTERFLIES	HOST PLANT
Snout butterfly (*Libytheana carinenta*)	Common hackberry (*Celtis occidentalis*)
Tawny emperor (*Asterocampa clyton*)	Common hackberry (*Celtis occidentalis*)
Variegated fritillary (*Euptoieta claudia*)	Violets (*Viola*)
Viceroy (*Limenitis archippus*)	Willows (*Salix*)
Coppers (Lycaenidae)	
American copper (*Lycaena phlaeas*)	Sheep sorrel (*Rumex acetosella*)
Elfins (Lycaenidae)	
Brown elfin (*Callophrys augustus*)	Blueberries (*Vaccinium*), Leather-leafs (*Chamaedaphne*)
Henry's elfin (*Callophrys henrici*)	Redbud (*Cercis canadensis*)
Pine elfin (*Callophrys niphon*)	Pitch pine (*Pinus rigida*), Virginia pine (*Pinus virginiana*), White pine (*Pinus strobus*)
Hairstreaks (Lycaenidae)	
Banded hairstreak (*Satyrium calanus*)	Bluejack oak (*Quercus incana*), Chestnut oak (*Quercus prinus*), Hickories (*Carya*), White oak (*Quercus alba*)
Coral hairstreak (*Harkencienus titus*)	Chickasaw plum (*Prunus angustifolia*), Chokecherry (*Prunus virginiana*), Wild cherry (*Prunus serotina*)
Gray hairstreak (*Strymon melinus*)	American vetch (*Vicia americana*), Bushclovers (*Lespedezia*), Tick trefoils (*Desmodium*)

(continued)

BUTTERFLIES	HOST PLANT
Hickory hairstreak (*Satyrium caryaevorum*)	Hickories (*Carya*)
Olive hairstreak (*Mitoura gruneus*)	Red cedar (*Juniperus virginiana*)
Red-banded hairstreak (*Calycopis cecrops*)	Oaks (*Quercus*), Staghorn sumac (*Rhus typhina*), Wax myrtle (*Myrica cerifera*), Winged sumac (*Rhus copallina*)
Striped hairstreak (*Satyrium liparops*)	American plum (*Prunus americana*), Blueberries (*Vaccinium*), Cherries (*Prunus*), Flame azalea (*Rhododendron calendulaceum*), Heath family (Ericaceae)
White M hairstreak (*Parrhasius m-album*)	Oaks (*Quercus*)
Harvesters (Lycaenidae)	
Harvester (*Feniseca tarquinius*)	Alders (*Alnus*) and beeches (*Fagus*) infested by woolly aphids (*Schizoneura* and *Pemphigus*)
Skippers (Hesperiidae)	
Common roadside skipper (*Amblyscirtes vialis*)	Spikegrass (*Uniola latifolia*)
Common sootywing (*Pholosara catullus*)	Strawberry blite (*Chenopodium capitatum*)
Crossline (*Polites origenes*)	Purpletop grass (*Tridens flavus*)
Dreamy duskywing (*Erynnis icelus*)	Aspens (*Populus*), Poplars (*Populus*), Willows (*Salix*)
Dun skipper (*Euphyes vestris*)	Sedges (*Carex*)
Dusted skipper (*Atrytonopsis hianna*)	Big bluestem (*Andropogon gerardii*), Little bluestem (*Schizachyrium scoparium*)

BUTTERFLIES	HOST PLANT
Hoary edge (*Achalarus lyciades*)	Devil's beggartick (*Bidens frondosa*), False indigo (*Amorpha fruticosa*), Hairy bushclover (*Lespedezia hirta*), Tick trefoils (*Desmodium*)
Hobomok skipper (*Poanes hobomok*)	Panicgrasses (*Panicum*)
Horace's duskywing (*Erynnis horatius*)	Oaks (*Quercus*)
Juvenal's duskywing (*Erynnis juvenalis*)	Oaks (*Quercus*)
Least skipper (*Ancyloxipha numitor*)	Rice cutgrass (*Leersia oryzoides*)
Little glassywing (*Pompeius verna*)	Purpletop grass (*Tridens flavus*)
Northern broken dash (*Wallengrenia egeremet*)	Cypress panicgrass (*Panicum dichotomum*), Deer-tongue grass (*Panicum clandestinum*), Fall panicgrass (*Panicum dichotomiflorum*)
Northern cloudywing (*Thorybes pylades*)	Legume family (Fabaceae)
Peck's skipper (*Polites peckius*)	Rice cutgrass (*Leersia oryzoides*)
Sachem (*Atalopedes campestris*)	Goosegrass (*Eleusine indica*), St. Augustine grass (*Stenotaphrum secundatum*)
Silver-spotted skipper (*Epargyreus clarus*)	False indigo (*Amorpha fruticosa*), Locusts and other legumes (*Robinia*)
Southern cloudywing (*Thorybes bathyllus*)	Legume family (Fabaceae)

(continued)

BUTTERFLIES	HOST PLANT
Swarthy skipper (*Nastra l'herminier*)	Little bluestem (*Schizachyrium scoparium*)
Tawny edged skipper (*Polites themistocles*)	Deer-tongue grass (*Panicum clandestinum*), Rice cutgrass (*Leersia oryzoides*), Switchgrass (*Panicum virgatum*)
Wild indigo duskywing (*Erynnis babtisiae*)	Blue false indigo (*Baptisia australis*), Wild indigo (*Baptisia tinctoria*)
Sulphurs (Pieridae)	
Clouded sulphur (*Colias philodice*)	Buffalo clover (*Trifolium stolo-niferum*)
Orange sulphur (*Colias eurytheme*)	Legume family (Fabaceae), Tick trefoils (*Desmodium*)
Swallowtails (Papilionidae)	
Black swallowtail (*Papilio polyxenes*)	Carrot family (Apiaceae)
Pipevine swallowtail (*Battus philenor*)	Dutchman's pipe (*Aristolochia durior*), Virginia snakeroot (*Aristolochia serpentaria*)
Spicebush swallowtail (*Papilio troilus*)	Sassafras (*Sassafras albidum*), Spicebush (*Lindera benzoin*)
Tiger swallowtail (*Papilio glaucus*)	Basswood (*Tilia americana*), Birches (*Betula*), Black cherry (*Prunus serotina*), Tulip poplar (*Liriodendron tulipifera*), Willows (*Salix*)
Zebra swallowtail (*Eurytides marcellus*)	Pawpaw (*Asimina triloba*)

BUTTERFLIES	HOST PLANT
Whites (Pieridae)	
Falcate orangetip (*Anthocharis midea*)	Crucifer family (Cruciferae), Rock cresses (*Arabis*)

MOTHS	HOST PLANT
Silk moths (Saturniidae)	
Cecropia moth (*Hyalophora cecropia*)	Black cherry (*Prunus serotina*), Flowering dogwood (*Cornus florida*), Gray birch (*Betula populifolia*), Oaks (*Quercus*), Poplars (*Populus*), Red maple (*Acer rubrum*), Sassafras (*Sassafras albidum*), Silver maple (*Acer saccharinum*)
Imperial moth (*Eacles imperialis*)	Basswood (*Tilia americana*), Birches (*Betula*), Black walnut (*Juglans nigra*), Elms (*Ulmus*), Maples (*Acer*), Oaks (*Quercus*), Pines (*Pinus*)
Io moth (*Automeris io*)	Beeches (*Fagus*), Cherries (*Prunus*), Maples (*Acer*), Oaks (*Quercus*), Poplars (*Populus*), Willows (*Salix*)
Larva: hickory horned devil	Black walnut (*Juglans nigra*), Persimmon (*Diospyros virginiana*), Sumacs (*Rhus*), Sweetgum (*Liquidambar styraciflua*)

(continued)

MOTHS	HOST PLANT
Luna moth (*Actias luna*)	Alders (*Alnus*), American beech (*Fagus grandifolia*), Birches (*Betula*), Hickories (*Carya*), Maples (*Acer*), Oaks (*Quercus*), Persimmon (*Diospyros virginiana*), Sweetgum (*Liquidambar styraciflua*), Willows (*Salix*)
Pink-striped oakworm (*Anisota virginiensis*)	Red oak (*Quercus rubra*)
Polyphemus moth (*Antheraea polyphemus*)	American chestnut (*Castanea dentata*), Basswood (*Tilia americana*), Birches (*Betula*), Elms (*Ulmus*), Hickories (*Carya*), Maples (*Acer*), Oaks (*Quercus*), Willows (*Salix*)
Promethea moth (*Callosamia promethea*)	Apple (*Pyrus malus*), Ashes (*Fraxinus*), Basswood (*Tilia americana*), Birches (*Betula*), Black cherry (*Prunus serotina*), Maples (*Acer*), Sassafras (*Sassafras albidum*), Spicebush (*Lindera benzoin*), Sweetgum (*Liquidambar styraciflua*), Tulip tree (*Liriodendron tulipifera*)
Rosy maple moth (*Dryocampa rubicunda*)	Maples (*Acer*)
Royal walnut moth (*Citheronia regalis*)	Hickories (*Carya*)
Sphinx moths (Sphingidae)	
Abbott's sphinx (*Sphecodina abbottii*)	Grapes (*Vitis*), Virginia creeper (*Parthenocissus quinquefolia*)

MOTHS	HOST PLANT
Apple sphinx (*Sphinx gordius*)	Ashes (*Fraxinus*), Carolina rose (*Rosa carolina*), Meadow rose (*Rosa blanda*), Southern crabapple (*Malus angustifolia*), Swamp rose (*Rosa palustris*)
Big poplar sphinx (*Pachysphinx modesta*)	Poplars (*Populus*), Willows (*Salix*)
Catalpa sphinx (*Ceratomia catalpae*)	Catalpa (*Catalpa bignonioides*)
Fawn sphinx (*Sphinx kalmiae*)	Ashes (*Fraxinus*), Fringe tree (*Chionanthus virginicus*), Poplars (*Populus*)
Four-horned sphinx (*Ceratomia amyntor*)	Birches (*Betula*), Elms (*Ulmus*)
Great ash sphinx (*Sphinx chersis*)	Ashes (*Fraxinus*)
Hermit sphinx (*Sphinx eremitus*)	Mints (Lamiaceae)
Huckleberry sphinx (*Paonias astylus*)	Blueberries (*Vaccinium*), Box huckleberry (*Gaylussacia brachycera*)
Hummingbird clearwing (*Hemaris thysbe*)	Hawthorns (*Crataegus*), Viburnums (*Viburnum*)
Pandora sphinx (*Eumorpha pandorus*)	Grapes (*Vitis*), Virginia creeper (*Parthenocissus quinquefolia*)
Pawpaw sphinx (*Dolba hyloeus*)	Pawpaw (*Asimina triloba*)
Pink-spotted hawkmoth (*Agrius cingulatus*)	Hedge false bindweed (*Calystegium sepium*)
Rustic sphinx (*Manduca rustica*)	Fringe tree (*Chionanthus virginicus*)
Small-eyed sphinx (*Paonias myops*)	Birches (*Betula*), Black cherry (*Prunus serotina*)

(continued)

MOTHS	HOST PLANT
Tobacco hornworm (*Manduca sexta*)	Horsenettle (*Solanum carolinense*)
Tomato hornworm (*Manduca quinquemaculata*)	Horsenettle (*Solanum carolinense*)
Trumpet vine sphinx (*Paratraea plebeja*)	Trumpet vine (*Campsis radicans*)
Virginia creeper sphinx (*Darapsa myron*)	Fox grape (*Vitis rotundifolia*), Virginia creeper (*Parthenocissus quinquefolia*)
Waved sphinx (*Ceratomia undulosa*)	Ashes (*Fraxinus*), Hawthorns (*Crataegus*), Oaks (*Quercus*)
White-lined sphinx (*Hyles lineata*)	Grapes (*Vitis*), Virginia creeper (*Parthenocissus quinquefolia*)
Wild cherry sphinx (*Sphinx drupiferarum*)	American plum (*Prunus americana*), Black cherry (*Prunus serotina*)
Other moths	
American dagger moth (*Acronicta americana*)	Basswood (*Tilia americana*), Maples (*Acer*), Oaks (*Quercus*), Willows (*Salix*)
Banded tussock moth (*Halysidota tessellaris*)	Ashes (*Fraxinus*), Black cherry (*Prunus serotina*), Box elder (*Acer negundo*), Elms (*Ulmus*)
Black-etched prominent (*Cerura scitiscripta*)	Black cherry (*Prunus serotina*), Poplars (*Populus*), Willows (*Salix*)
Dogbane tiger moth (*Cycnia tenera*)	Dogbanes (*Apocynum*)
Eastern tent caterpillar (*Malacosoma americanum*)	Black cherry (*Prunus serotina*)

MOTHS	HOST PLANT
Giant leopard moth (*Hypercompe scribonia*)	Maples (*Acer*), Violets (*Viola*), Willows (*Salix*)
Large tolype (*Tolype velleda*)	Apples (*Malus*), Ashes (*Fraxinus*), Elms (*Ulmus*), Oaks (*Quercus*), Willows (*Salix*)
Major datana (*Datana major*)	Azaleas (*Rhododendron*), Blueberries (*Vaccinium*)
Milkweed tussock moth (*Euchaetes egle*)	Milkweeds, butterfly weeds (*Asclepias*)
Saddleback caterpillar moth (*Acharia stimulea*)	Asters (Asteraceae), Blueberries (*Vaccinium*), Cherries (*Prunus*), Oaks (*Quercus*)
Saltmarsh moth (*Estigmene acrea*)	Cordgrass (*Spartina alterniflora*)
Spotted apatelodes (*Apatelodes torrefacta*)	Black cherry (*Prunus serotina*), Maples (*Acer*), Oaks (*Quercus*)
Virginia ctenucha (*Ctenucha virginica*)	Grass family (Poaceae), Irises (*Iris*), Rushes (*Juncus*)
Virginia tiger moth (*Spilosoma virginica*)	Birches (*Betula*), Maples (*Acer*)
Virgin tiger moth (*Grammia virgo*)	Bedstraws (*Galium*)
White-marked tussock moth (*Orgyia leucostigma*)	American sycamore (*Platanus occidentalis*), Apples (*Malus*), Basswood (*Tilia americana*), Birches (*Betula*), Elms (*Ulmus*), Poplars (*Populus*)
Yellow-necked caterpillar (*Datana ministra*)	Black cherry (*Prunus serotina*), Elms (*Ulmus*), Maples (*Acer*), Oaks (*Quercus*), Walnut (*Juglans*)

APPENDIX THREE

Experimental Evidence

In a survey of insect herbivores found eating woody native and alien species in Oxford, Pennsylvania, native plants produced over four times more insect biomass than alien plants produced. This difference resulted entirely from the inability of insects with chewing mouthparts to eat alien plants (Tallamy, unpublished data).

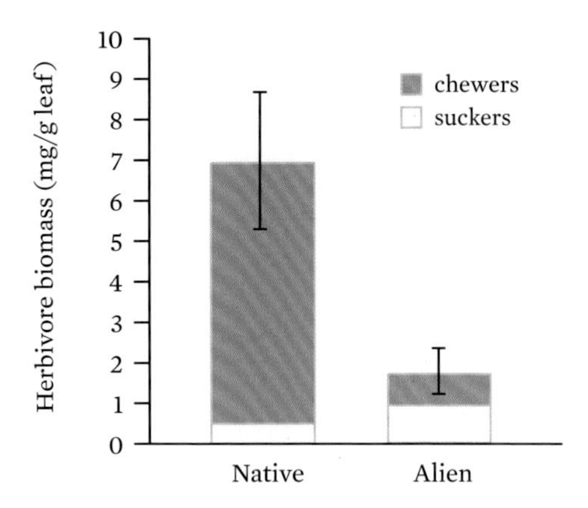

In a comparison of the diversity of herbivorous insects on native and alien woody plants in Oxford, Pennsylvania, more than three times as many insect species were associated with native plants as with alien plants (Tallamy, unpublished data).

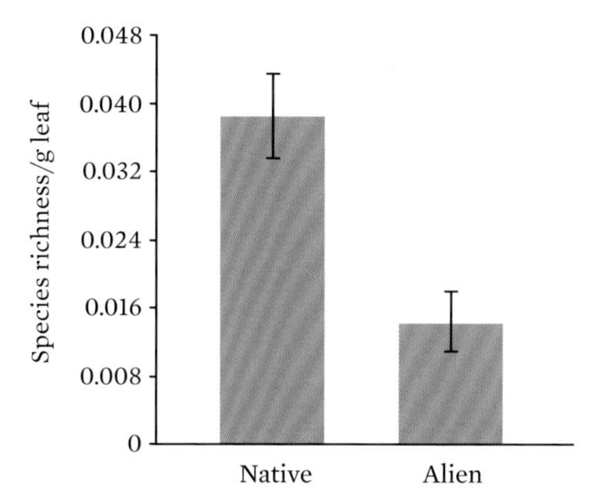

In a comparison of Lepidoptera larvae produced on native and alien woody plants in Oxford, Pennsylvania, native plants supported 35 times more caterpillar biomass, the preferred source of protein for most bird nestlings, than alien plants supported (Tallamy, unpublished data).

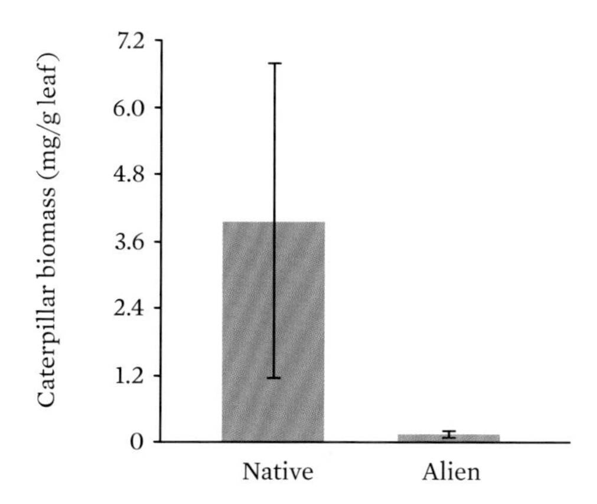

In a survey of generalist insect species on native and alien woody plants in Oxford, Pennsylvania, twice as much biomass of generalist insect species is produced on native plants as on aliens. This suggests that even generalists (insects capable of eating several plant species) are not able to eat and grow on alien plants as well as they do on natives (Tallamy, unpublished data).

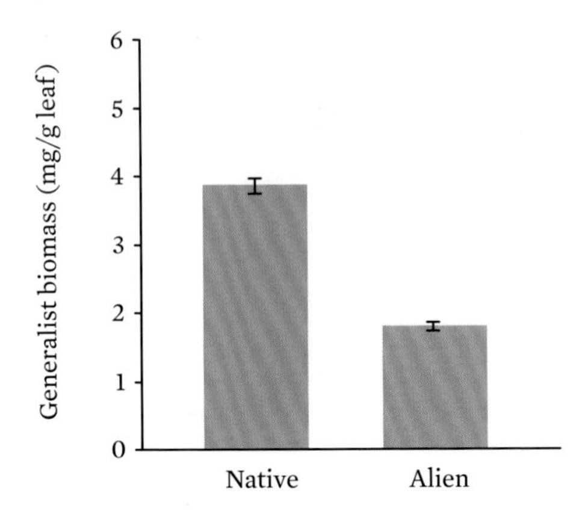

In a comparison of the generalist insect fauna on early successional perennials in Newark, Delaware (Ballard 2006), insect biomass produced by native plants was significantly greater than that produced by alien plants.

TOTAL NATIVE INSECTS	JULY 2004	AUGUST 2004	JULY 2005	AUGUST 2005
Native plants	30.7 (6.3)	90.7 (25.2)	96.4 (25.1)	398.3 (85.7)
Alien plants	14.0 (4.2)	40.9 (10.3)	22.8 (8.4)	70.9 (13.4)
P	0.04	0.20	.05	0.02
NATIVE HERBIVORES	**JULY 2004**	**AUGUST 2004**	**JULY 2005**	**AUGUST 2005**
Native plants	18.5 (3.1)	36.1 (14.8)	82.2 (22.0)	373.3 (84.0)
Alien plants	9.6 (2.5)	20.5 (10.4)	16.4 (4.1)	67.4 (13.8)
P	0.10	0.10	0.05	0.02

Statistical tests performed on log-transformed data for July and August 2005; untransformed means and SEM reported (mean [SEM] mg/100g dry leaf biomass).

REFERENCES

Amman, G. D., and C. F. Speers. 1965. Balsam woolly aphid in the southern Appalachians. *Journal of Forestry* 63: 18–20.

Anagnostakis, S. L. 1987. Chestnut blight: The classical problem of an introduced pathogen. *Mycologia* 29: 23–37.

Annecke, D. P., and V. C. Moran. 1978. Critical reviews of biological pest control in South Africa. Part 2, The prickly pear, *Opuntia ficus-indica* (L.) Miller. *Journal of the Entomological Society of Southern Africa* 41: 161–188.

Arnett, R. H. 2000. *American Insects: A Handbook of the Insects of America North of Mexico*. Boca Raton, Florida: CRC Press.

Austin, D. 1978. Exotic plants and their effects on southeastern Florida. *Environmental Conservation* 5: 230–234.

Ballard, M. 2006. Insect populations on early successional native and alien plants. Master's thesis, University of Delaware, Newark.

Bailey J. P, L. E. Child, L.C. de Waal, and M. Wade. 1995. The invasive nature of *Fallopia japonica* is enhanced by vegetative regeneration from stem tissues. In *Plant Invasions:*

General Aspects and Special Problems, edited by P. Pysek, K. Prach, M. Rejmanek, and M. Wade, 131–139. Amsterdam: SPB Academic.

Bailey, R. 1993. *Ecoscam: The False Prophets of Ecological Apocalypse*. New York: Saint Martin's Press.

Berenbaum, M. R. 1990. Coevolution between herbivorous insects and plants: Tempo and orchestration. In *Insect Life Cycles*, edited by F. Gilbert, 87–89. London: Springer.

Bernays, E. M., and M. Graham. 1988. On the evolution of host specificity in phytophagous arthropods. *Ecology* 69: 886–892.

Brattsten, L. B., C. F. Wilkinson, and T. Eisner. 1977. Herbivore-plant interactions: Mixed function oxidases and secondary plant substances. *Science* 196: 1349–1352.

Brooks, M. L., C. M. D'Antonio, D. M. Richardson, J. B. Grace, J. E. Keely, J. M. DiTomaso, R.

J. Hobbs, M. Pellant, and D. Pyke. 2004. Effects of invasive alien plants on fire regimes. *BioScience* 54: 677–688.

Brower, L. P. 1995. Understanding and misunderstanding the migration of the monarch butterfly (Nymphalidae) in North America, 1857–1995. *Journal of the Lepidopterists' Society* 49: 304–385.

Brower, L. P., D. R. Kust, E. Rendon-Salinas, E. G. Serrano, K. R. Kust, J. Miller, C. Fernandez del Rey, and K. Pape. 2004. Catastrophic winter storm mortality of monarch butterflies in Mexico in January 2002. In *Monarch Butterfly Biology and Conservation*. Ithaca: Cornell University Press.

Brown, W. P. 2006. On the community composition and abundance of Delaware forest birds. Ph.D. diss., University of Delaware, Newark.

Butler, J. L., and D. R. Cogan. 2004. Leafy spurge effects on patterns of plant species richness. *Journal of Range Management* 57: 305–311.

Cech, F. C. 1986. American chestnut (*Castanea dentata*): Replacement species and current status. *In Endangered and Threatened Species Programs in Pennsylvania and Other States: Causes, Issues, and Management*, edited by S. K. Majumdar, F. J. Brenner, and A. F. Rhoads, 145–155. Easton: Pennsylvania Academy of Science.

Cerqueira, N. 2005. Pollinator visitation preference on native and non-native congeneric plants. Master's thesis, University of Delaware, Newark.

Cocroft, R. B. 1999. Offspring-parent communication in a subsocial treehopper (Hemiptera: Membracidae: *Umbonia crassicornis*). *Behaviour* 136: 1–21.

Coleman, D. 2003. Pennsylvania's wild areas. *Sylvanian*, spring, 11.

Conrad, K. F., M. S. Warren, R. Fox, M. S. Parsons, and I. P. Woiwod. 2006. Rapid declines of common, widespread British moths provide evidence of an insect biodiversity crisis. *Biological Conservation* 132: 2779–2791.

Cowles, R. S. 2003. Modeling the effectiveness of bifenthrin for protecting container-grown crops from Japanese and oriental beetle larvae. *Journal of Environmental Horticulture* 21: 78–81.

Costello, S. L., P. D. Pratt, M. B. Rayamajhi, and T. D. Center. 1995. Arthropods associated with above ground portions of the invasive tree, *Melaleuca quinquenervia,* in South Florida, USA. *Florida Entomologist* 86: 300–322.

Crawley, M. J. 1989. The relative importance of vertebrate and invertebrate herbivores in plant population dynamics. In *Insect-Plant Interactions,* edited by E. A. Bernays, 45–71. Boca Raton, Florida: CRC Press.

De Ruiter, P. C., V. Wolters, J. C. Moore, and K. O. Winemiller. 2005. Food web ecology: Playing Jenga and beyond. *Science* 309: 68–71.

Diamond, S. J., R. H. Giles Jr., R. L Kirkpatrick, and G. J. Griffin. 2000. Hard mast production before and after the chestnut blight. *Southern Journal of Applied Forestry* 24: 196–201.

Dickinson, M. B. 1999. *Field Guide to the Birds of North America*. 3rd ed. Washington, D.C.: National Geographic Society.

Dingle, H. 1972. Migration strategies of insects. *Science* 175: 1327–1335.

Dirr, M. 1998. *Manual of Woody Landscape Plants*. Champaign, Illinois: Stipes.

Dixon, A. F. G. 1997. *Aphid Ecology: An Optimization Approach*. 2nd ed. New York: Springer.

Dobson A. P. 1996. *Conservation and Biodiversity*. New York: Scientific American Library.

Duguay, J. P., P. B. Wood, and G. W. Miller. 2000. Effects of timer harvests on invertebrate biomass and avian nest success. *Wildlife Society Bulletin* 28: 1123–1131.

Duncan, C. 1997. *Environmental Benefits of Weed Management*. Washington, D.C.: Dow Elanco.

Ehrlich, P. R., and A. H. Ehrlich. 1981. *Extinction: The Causes and Consequences of the Disappearance of Species*. New York: Random House.

Ehrlich, P. R., and P. H. Raven. 1965. Butterflies and plants: A study in coevolution. *Evolution* 19: 586–608.

Elton, C. 1927. *Animal Ecology*. New York: Macmillan.

Elvidge, C. D., C. Milesi, J. B. Dietz, B. T. Tuttle, P. C. Sutton, R. Nemani, and J. E. Vogelmann. 2004. U.S. constructed area approaches the size of Ohio. *Eos: Transactions of the American Geophysical Union* 85 (24): 233.

Farrell, B., and C. Mitter. 1990. Phylogenesis of insect/plant interactions: Have the *Phyllobrotica* leaf beetles (Chrysomelidae) and the Lamiales diversified in parallel? *Evolution* 44: 1389–1403.

Flanders, A. A., W. P. Kuvlesky Jr., D. C. Ruthven III, R. E. Zaiglin, R. L. Bingham, T. E. Fulbright, F. Hernandez, and L. A. Brennan. 2006. Effects of invasive exotic grasses on South Texas rangeland breeding birds. *The Auk* 123: 171–182.

Funk, D. H. 1989. The mating of tree crickets. *Scientific American* 260: 50–59.

Futuyma, D. J., and F. Gould. 1979. Associations of plants and insects in a deciduous forest. *Ecological Monographs* 49: 33–50.

Garbelotto, M. 2004. Sudden oak death: A tale of two continents. *Outlooks on Pest Management*, April, 85–89.

Goold, C. 1994. The high cost of weeds. In *Noxious Weeds: Changing the Face of Southwestern Colorado*, 5–6. Durango, Colorado: San Juan National Forest Association.

Gordon, D. R. 1998. Effects of invasive, non-indigenous plant species on ecosystem processes: Lessons from Florida. *Ecological Applications* 8: 975–989. Graves, A. H. 1930. Progress toward the development of disease resistant strains of chestnut. *Brooklyn Botanic Garden Record* 19: 62–67.

Grey, G. W., and F. J. Deneke. 1986. *Urban Forestry*. 2nd ed. Toronto: Wiley.

Hansens, E. J., and H. B. Weiss. 1954. Entomology in New Jersey. Rutgers Entomology. http://www.rci.rutgers.edu/~insects/hist.htm.

Hay, M., and P. Steinberg. 1992. The chemical ecology of plant-herbivore interactions in marine versus terrestrial communities. In *Herbivores: Their Interactions with Secondary Metabolites*, edited by G. Rosenthal and M. Berenbaum, 371–413. Evolutionary and Ecological Processes. San Diego: Academic Press.

Hayden, D. 2004. *A Field Guide to Sprawl*. New York: Norton.

Heinrich, B., and S. L. Collins. 1983. Caterpillar leaf damage and the game of hide-and-seek with birds. *Ecology* 64: 592–602.

Hess, G. K., R. L. West, M. V. Barnhill, and L. M. Fleming. 2000. *Birds of Delaware*. Pittsburgh: University of Pittsburgh Press.

Hobbs, R. J., and S. E. Humphries. 1995. An integrated approach to the ecology and management of plant invasions. *Conservation Biology* 9: 761–770.

Hokkanen, H. M. T., and D. Pimentel. 1989. New associations in biological control—theory and practice. *Canadian Entomologist*. 121: 829–840.

Holmes, H. 2006. *Suburban Safari: A Year on the Lawn*. New York: Bloomsbury.

Houston, D. R., E. J. Parker, R. Perrin, and K. J. Lang. 1979. Beech bark disease: A comparison of the disease in North America, Great Britain, France and Germany. *European Journal of Forest Pathology*. 9: 199–211.

Hutchinson, J. M. S., N. J. Laethers, J. Herynk, L. R. Campbell, and J. C. Reese. 2003 Agricultural plant pathogen disease pathways: Predicting the dispersal of exotic soybean aphids. In *Papers of the Applied Geography Conferences*, edited by G. A. Tobin and B. E. Montz, 26: 471–478. Colorado Springs, Colorado.

Ibarra-F., F. A., J. R. Cox, M. H. Martin-R., T. A. Crowl, and C. A. Call. 1995. Predicting buffelgrass survival across a geographical and environmental gradient. *Journal of Range Management* 48: 53–59.

Jaques, H. E. 1918. A long-life wood-boring beetle. *Proceedings of the Iowa Academy of Science*. 25: 175.

Japanese Knotweed Alliance. 1999. http://www.cabi-bioscience.org/html/japanese_knotweed_alliance.htm.

Jenkins, V. S. 1994. *The Lawn: A History of an American Obsession*. Washington, D.C.: Smithsonian Institution Press.

Kellert, S. R., and E. O. Wilson, eds. 1993. *The Biophylia Hypothesis*. Washington, D.C.: Island Press.

Kennedy, C. E. J., and T. R. E. Southwood. 1984. The number of insects associated with British trees: A re-analysis. *Animal Ecology* 53: 455–478.

Kennedy, D., and B. Hanson. 2006. Ice and history. *Science* 311: 1673.

Kennedy, T. A., S. Naeem, K. M. Howe, J. M. H. Knops, D. Tilman, and P. Reich. 2002. Biodiversity as a barrier to ecological invasion. *Nature* 417: 636–638.

Kingsbury, J. M. 1964. *Poisonous Plants of the United States and Canada*. Englewood Cliffs, New Jersey: Prentice-Hall.

Kinzig, A. P., S. W. Pacala, and D. Tilman. 2002. *Functional Consequences of Biodiversity: Empirical Progress and Theoretical Extensions*. Princeton: Princeton University Press.

Kirby, D. 2004. Threat of sudden oak death grows in Georgia. *Georgia Outdoor News*. http://www.gon.com/oak704.html.

Klem, D. Jr. 1990. Collisions between birds and windows: Mortality and prevention. *Journal of Field Ornithology* 61: 120–128.

Leopold, D. J. 2005. *Native Plants of the Northeast: A Guide for Gardening and Conservation*. Portland, Oregon: Timber Press.

Lesica, P., and T. H. DeLuca. 1996. Long-term harmful effects of crested wheatgrass on Great Plains grassland ecosystems. *Journal of Soil and Water Conservation* 51: 408–409.

Lill, J. T., and R. J. Marquis. 2003. Ecosystem engineering by caterpillars increases insect herbivore diversity on white oak. *Ecology* 84: 682–690.

Lloyd, J. D., and T. E. Martin. 2005. Reproductive success of chestnut-collared longspurs in native and exotic grassland. *Condor* 107: 363–374.

Losey, J., and M. Vaughan. 2006. The economic value of ecological services provided by insects. *BioScience* 56: 311–323.

Lyon, W. F. 1996. Insects as human food. Ohio State University Extension Fact Sheet. HYG 2160-96.

MacArthur, R. H., and E. O. Wilson. 1967. *The Theory of Island Biogeography*. Princeton: Princeton University Press.

Macfarlane, R. P., and H. J. van den Ende. 1995. Vine-feeding insects of oldman's beard, *Clematis vitalba*, in New Zealand. In *Proceedings of the Eighth International Symposium on Biological Control of Weeds, 2–7 February 1992, Lincoln University, Canterbury, New Zealand*, edited by E. S. Deltosse and R. R. Scott, 57–58. Melbourne: DSIR/CSIRO.

Mack, R. N., and M. Erneberg. 2002. The United States naturalized flora: Largely the product of deliberate introductions. *Annals of the Missouri Botanical Gardens* 89: 176–189.

Mack, R. N., D. Simberloff, W. M. Lonsdale, H. Evans, M. Clout, and F. A. Bazzaz. 2000. Biotic invasion: Causes, epidemiology, global consequences, and control. *Ecological Applications*. 10: 689–710.

Marra, P. P., K. A. Hobson, and R. T. Holmes. 1998. Linking winter and summer events in a migratory bird by using stable-carbon isotopes. *Science* 282: 1884–1886.

Marren, P. 2001. What time hath stole away: Local extinctions in our native flora. *British Wildlife*, June, 305–310.

May, R. M. 1973. *Stability and Complexity in Model Ecosystems*. 2nd ed. Princeton: Princeton University Press.

Mayr, E. 1942. *Systematics and the Origin of Species*. New York: Columbia University Press.

McClure, M. S., S. M. Salom, and K. S. Shields. 2001. Hemlock woolly adelgid. Report by the USDA Forest Health Technology Enterprise Team. http://www.fs.fed.us/na/morgantown/fhp/hwa/pub/fhtet-2001-03.pdf.

McKinney, M. L. 2002. Urbanization, biodiversity, and conservation. *BioScience* 52: 883–890.

Milesi, C., S. W. Running, C. D. Elvidge, J. B. Dietz, B. T. Tuttle, and R. R. Nemani. 2005. Mapping and modeling the biogeochemical cycling of turf grasses in the United States. *Environmental Management* 36 (3): 426–438.

Miller, J. A. 2003. *Nonnative Invasive Plants of Southern Forests: A Field Guide for Identification and Control*. Gen. Tech. Rep. SRS–62. Asheville, N.C.: USDA Forest Service, Southern Research Station.

Mooney, H. A., and R. J. Hobbs, eds. 2000. *Invasive Species in a Changing World*. Washington, D.C.: Island Press

Moore, J. C., E. L. Berlow, D. C. Coleman, P. C. de Ruiter, Q. Dong, A. Hastings, N. Collins Johnson, K. S. McCann, K. Melville, P. J. Morin, K. Nadelhoffer, A. D. Rosemond, D. M. Post, J. L. Sabo, K. M. Scow, M. J. Vanni, and D. H. Wall. 2004. Detritus, trophic dynamics, and biodiversity. *Ecology Letters* 7: 584–600.

Morrow, P. A., and V. C. LaMarche Jr. 1978. Tree ring evidence for chronic insect suppression of productivity in subalpine eucalyptus. *Science* 201: 1244–1246.

Moul, E. T. 1948. A dangerous polygonum in Pennsylvania. *Rhodora* 50: 64–66.

National Academies. 2005. *Understanding and Responding to Climate Change: Highlights of National Academies Reports*. http://dels.nas.edu/basc/Climate-HIGH.pdf

Nielson, G. R. 1989. *Asiatic Garden Beetle*. Extension leaflet 247. University of Vermont Extension, Department of Plant and Soil Science.

——. 1992. *European Chafer*. Extension leaflet 199. University of Vermont Extension, Department of Plant and Soil Science.

Nott, M. P., D. F. Desante, R. B. Siegel, and P. Pyle. 2002. Influences of the El Nino/Southern Oscillation and the North Atlantic Oscillation on avian productivity in forests of the Pacific Northwest of North America. *Global Ecology and Biogeography* 11: 333–342.

Novotny, V., P. Drozd, S. E. Miller, M. Kulfan, M. Janda, Y. Basset, and G. D. Weiblen. 2006. Why are there so many species of herbivorous insects in the tropical rainforests. *Science* 313: 1115–1118.

Nowak, D. J., and A. R. Rowan. 1990. History and range of Norway maple. *Journal of Arboriculture* 16: 291–296.

Opler, P. A., and V. Malikul. 1992. *A Field Guide to Eastern Butterflies*. New York: Houghton Mifflin.

Ostry, M. E, S. A. Katovich, and R. L. Anderson. 1997. First report of *Sirococcus clavigignenti-juglandacearum* on black walnut. *Plant Disease* 81: 830.

Paine, R. T. 1969. The Pisaster-Tegula interaction: Prey patches, predator food preference, and individual community structure. *Ecology* 50: 950–961.

——. 1980. Food webs: Linkage, interaction strength and community infrastructure. *Journal of Animal Ecology* 49: 667–685.

Palm, M. E. 2001. Systematics and the impact of invasive fungi on agriculture in the United States. *BioScience* 51: 141–147.

Parker, J. D., and M. E. Hay. 2005. Biotic resistance to plant invasions? Native herbivores prefer non-native plants. *Ecology Letters* 8: 959–967.

Parker, J. D., D. E. Burkepile, and M. E. Hay. 2006. Opposing effects of native and exotic herbivores on plant invasions. *Science* 311: 1459–1461.

Powell, G. H. 1900. The European and Japanese chestnuts in the eastern United States. *In Eleventh Annual Report of the Delaware College Agricultural Experiment Station*, 101–135. Newark, Delaware.

Power, M. E., D. Tilman, J. A. Estes, B. A., Menge, W. J. Bond, L. S. Mills, G. Daily, J. C. Castilla, J. Lubchenco, and R. T. Paine. 1996. Challenges in the quest for keystones. *Bioscience* 46: 610–620.

Ragsdale, D. W., D. J. Voegtlin, and R. J. O'Neil. 2004. Soybean aphid biology in North America. *Annals of the Entomological Society of America* 97: 204–208.

Randall, J. 1996. Weed control for the preservation of biological diversity. *Weed Technology* 10: 370–383.

Rasor, Y. 1999. Not so fly. *Feminista!* 3: 3.

Raupp, M. J., and R. M. Noland. 1984. Implementing landscape plant management programs in institutional and residential settings. *Journal of Arboriculture*. 10: 161–169.

Ray, D. L., and L. R. Guzzo. 1994. *Environmental Overkill: Whatever Happened To Common Sense?* Washington, D.C.: Harper Collins.

Reichard, S. H., and P. White. 2001. Horticulture as a pathway of invasive plant introductions in the United States. *BioScience* 51: 103–113.

Rhoads, A. F., and T. A. Block. 2005. *Trees of Pennsylvania*. Philadelphia: University of Pennsylvania Press.

Ritland, D. B. and L. P. Brower. 1991. The viceroy butterfly is not a Batesian mimic. *Nature* 350: 497–498.

Robinson, G. S., P. R. Ackery, I. J. Kitching, G. W. Beccaloni, and L. M. Hernandez. 2002. *Host Plants of the Moth and Butterfly Caterpillars of America North of Mexico*. Memoirs of the American Entomological Institute, vol. 69. Gainesville, Florida.

Robinson, W. D. 1999. Long-term changes in the avifauna of Barro Colorado Island, Panama: A tropical forest Isolate. *Conservation Biology* 13: 85–97.

Robson, M. 2001. Citrus longhorned beetle: Get acquainted here. Gardening in Western Washington. Washington State University Extension Service. http://gardening.wsu.edu/column/08-26-01.htm.

Rosenthal, G. A., and D. H. Janzen, eds. 1979. *Herbivores: Their Interaction with Secondary Plant Metabolites*. New York: Academic Press.

Rosenzweig, M. L. 1995. *Species Diversity in Space and Time*. New York: Cambridge University Press.

——. 2003. *Win-Win Ecology: How the Earth's Species Can Survive in the Midst of Human Enterprise*. New York: Oxford University Press.

Ruby, E. G., C. O. Wirsen, and H. W. Jannasch. 1981. Chemolithotrophic sulfur-oxidizing bacteria from the Galapagos Rift hydrothermal vents. *Applied and Environmental Microbiology* 42: 31–324.

Sadof, C. S., and M. J. Raupp. 1996. Aesthetic thresholds and their development. In *Economic Thresholds for Integrated Pest Management*, edited by L. G. Higley and L. P. Pedigo, 203–226. Lincoln: University of Nebraska Press.

Sauer, J. R., J. E. Hines, and J. Fallon. 2005. *The North American Breeding Bird Survey: Results and Analysis, 1966–2004*. Version 2005.2. Laurel, Maryland: USGS Patuxent Wildlife Research Center.

Sekercioglu, C., G. C. Daily, and P. R. Ehrlich. 2004. Ecosystem consequences of bird declines. *Proceedings of the National Academy of Sciences* 101: 18042–18047.

Shrewsbury, P. M., and M. J. Raupp. 2006. Do top-down or bottom-up forces determine

Stephanitis pyrioides abundance in urban landscapes? *Ecological Applications* 16: 262–272.

Silvertown, J. W. *Demons in Eden: The Paradox of Plant Diversity*. Chicago: University of Chicago Press.

Sinclair, W. A., and H. M. Griffiths. 1994. Ash yellows and its relationship to dieback and decline of ash. *Annual Review of Phytopathology* 32: 49–60.

Smith, D. M. 2000. American chestnut: Ill-fated monarch of the eastern hardwood forest. *Journal of Forestry*, February, 12–15.

Stein, S. 1995. We don't garden right. *Green Scene*, September, 3–6.

Sternberg, G., and J. Wilson. 2004. *Native Trees for North American Landscapes: From the Atlantic to the Rockies*. Portland, Oregon: Timber Press.

Stokstad, E. 2006. New disease endangers Florida's already-suffering citrus trees. *Science* 312: 523–524.

Strong, D. R., J. H. Lawton, and R. Southwood. 1984. *Insects on Plants: Community Patterns and Mechanisms*. Cambridge: Harvard University Press.

Tallamy, D. W. 2004. Do alien plants reduce insect biomass? *Conservation Biology* 18: 1–4.

Tallamy, D. W., and L. A. Horton. 1990. Costs and benefits of the egg dumping alternative in *Gargaphia* lace bugs (Hemiptera: Tingidae). *Animal Behaviour* 39: 52–360.

Tallamy, D. W., and T. K. Wood. 1986. Convergence patterns in subsocial insects. *Annual Review of Entomology* 31: 369–390.

Tewksbury, L., R. Casagrande, B. Bloosey, P. Häfliger, M. Schwarzländer. 2002. Potential for biological control of *Phragmites australis* in north America. *Biological Control* 23: 191–212.

They paved paradise. 2004. *OnEarth*, fall, 7.

Tilman, D. 2000. Causes, consequences and ethics of biodiversity. *Nature* 405: 208–211.

Tilman, D., D. Wedin, and J. Knops. 1996. Productivity and sustainability influenced by biodiversity in grassland ecosystems. *Nature* 379: 718–720.

Triplehorn, C. A., and N. F. Johnson. 2005. *Borror and DeLong's Introduction to the Study of Insects*. 7th ed. Belmont, California: Thomson Brooks/Cole.

Tyser, R. 1992. Vegetation associated with two alien plant species in a fescue grassland in Glacier National Park, Montana. *Great Basin Naturalist* 52: 193–198.

U.S. Census Bureau. 2005. http://www.census.gov.

USDA Census of Agriculture. 2002. http://www.nass.usda.gov/census/.

U.S. Department of Energy, Energy Information Administration. 1995. *Emissions of Greenhouse*

Gases in the United States, 1987–1994. Pittsburgh: U.S. Government Printing Office.

U.S. Fish & Wildlife Service. 2002. Migratory bird mortality.http://www.fws.gov/birds/mortality-fact-sheet.pdf.

U.S. Forest Service. 2005. Emerald ash borer. http://www.na.fs.fed.us/fhp/eab/. Valencia, R., H. Balslev, and G. Paz y Mino. 1994. High tree alpha-diversity in Amazonian Ecuador. *Biodiversity and Conservation* 3: 21–28. Van Mantgem, P. J., N. L. Stephenson, M. B. Keifer, and J. E. Keeley. 2004. Effects of an introduced pathogen and fire exclusion on the demography of sugar pine. *Ecological Applications* 14: 1590–1602.

Vargo, B. D., and S. Gallagher. 2002. Dirty little secrets. *Delaware Today*, May, 62–69.

Wagner, D. L. 2005. *Caterpillars of Eastern North America*. Princeton: Princeton University Press.

Waterworth, H. E., and G. A White. 1982. Plant introductions and quarantine: The need for both. *Plant Disease* 66: 87–90.

Weiss, A., and M. R. Berenbaum. 1988. Herbivorous insects/plant interactions. In *Plant-Animal Interactions—a Textbook*, edited by W. G. Abrahamson, 140–183. New York: Macmillan.

Wilcove, D. S., D. Rothstein, J. Dubow, A. Philips, and E. Losos. 1998. Quantifying threats to imperiled species in the United States. *BioScience* 48: 607–615.

Williamson, M. 1996. *Biological Invasions*. London: Chapman & Hall.

Willis, K. J., and J. C. McElwain. 2002. *The Evolution of Plants*. Oxford: Oxford University Press.

Wilson, E. O. 1975. Sociobiology: The New Synthesis. Cambridge: Harvard University Press.

———. 1987. The little things that run the world: The importance and conservation of invertebrates. *Conservation Biology* 1: 344–346.

———. 1999. *The Diversity of Life*. New York: Norton.

———. 2002. *Biophilia*. Cambridge: Harvard University Press.

Winter, T. G. 1974. New host plant records of Lepidoptera associated with conifer afforestation in Britain. *Entomologist's Gazette* 25: 247–258.

Yeargan, K. V. 1988. Ecology of a bolas spider, *Mastophora hutchinsoni*: Phenology, hunting tactics, and evidence for aggressive chemical mimicry. *Oecologia* 74: 524–530.

Zanette, P., P. Doyle, and S. M. Tremont. 2000. Food shortage in small fragments: Evidence from an area-sensitive passerine. *Ecology* 81: 1654–1666.

INDEX

ABOUT THE AUTHOR

Lisa Mattei / New England Wild Flower Society

Douglas W. Tallamy is professor and chair of the Department of Entomology and Wildlife Ecology at the University of Delaware in Newark, Delaware, where he has published 68 research articles and has taught insect taxonomy, behavioral ecology, and other courses for nearly three decades. Chief among his research goals is to better understand the many ways insects interact with plants and how such interactions determine the diversity of animal communities. The Garden Writers Association of America awarded *Bringing Nature Home* its Silver Medal in 2008.